The theorems of Hawking and Penrose show that space-times are likely to contain incomplete geodesics. Such geodesics are said to end at a singularity if it is impossible to continue the space-time and geodesic without violating the usual topological and smoothness conditions on the space-time. In this book the different possible singularities are defined, and the mathematical methods needed to extend the space-time are described in detail. The results obtained (many appearing here for the first time) show that singularities are associated with a lack of smoothness in the Riemann tensor. While the Friedmann singularity is analysed as an example, the emphasis is on general theorems and techniques rather than on the classification of particular exact solutions.

Cambridge Lecture Notes in Physics 1

General Editors: P. Goddard, J. Yeomans

The Analysis of Space-Time Singularities

Cambridge Lecture Notes in Physics

The Analysis of Space-Time Singularities

C. J. S. CLARKE

University of Southampton

CAMBRIDGE
UNIVERSITY PRESS

Published by the Press Syndicate of the University of Cambridge
The Pitt Building, Trumpington Street, Cambridge CB2 1RP
40 West 20th Street, New York, NY 10011–4211, USA
10 Stamford Road, Oakleigh, Melbourne 3166, Australia

First published 1993

Printed in Great Britain at the University Press, Cambridge

A catalogue record for this book is available from the British Library

Library of Congress cataloguing in publication data applied for

ISBN 0 521 43796 2 paperback

Contents

Preface

The central aim of this book is the development of the results and techniques needed to determine when it is possible to extend a space-time through an "apparent singularity" (meaning, a boundary-point associated with some sort of incompleteness in the space-time). Having achieved this, we shall obtain a characterisation of a "genuine singularity" as a place where such an extension is not possible. Thus we are proceeding by elimination: rather than embarking on a direct study of genuine singularities, we study extensions in order to rule out all apparent singularities that are not genuine. It will turn out, roughly speaking, that the genuine singularities which then remain are associated either with some sort of topological obstruction to the construction of an extension, or with the unboundedness of the Riemann tensor when its size is measured in a suitable norm.

I had at one stage hoped that there would be a single simple criterion for when such an extension cannot be constructed, which would then lay down once and for all what a genuine singularity is. But it seems that this is not to be had: instead one has a variety of possible tools and concepts for constructing extensions, and when these fail one declares the space-time to be singular on pragmatic grounds. The main such tools are the use of Hölder and Sobolev norms of functions, used for measuring the extent to which the metric or the Riemann tensor is irregular.

The introduction of Sobolev norms, in addition to Hölder ones, is an attempt to carry the work in this book beyond that in my previous publications (see Clarke, 1982). The need for this is twofold: first, the use of this norm enables one to make a connection with the existence theorems for Einstein's equations, which are all formulated in terms of Sobolev spaces. Second, the requirement that a Sobolev norm be definable places a much weaker condition on the Riemann tensor than is the case with Hölder norms or differentiability conditions in the pointwise sense, the Riemann tensor not even having to be bounded in the case of the Sobolev norms relevant to general relativity. Thus the failure of the Sobolev norm at a genuine singularity implies that the Riemann tensor has there a decidedly worse behaviour than is the case with a failure of Hölder continuity.

The penalty paid for this potential improvement is the technical difficulty of linking differential geometry with Sobolev norms on the Riemann tensor, a difficulty compounded by the fact that most relativists are unfamiliar with the techniques of real analysis used in the Sobolev space theory. I shall therefore be devoting some space in the book to explaining these techniques and sketching those key results in the theory that show how the Sobolev spaces enter into existence theorems and singularity theory. Some of these difficulties remain to be overcome, so that it has not proved possible to give a complete treatment using only Sobolev norms on the Riemann tensor. At present the situation appears to be that one can work either with Hölder norms on the Riemann tensor, or Sobolev norms on the metric components. But I have tried to give the Sobolev versions of results wherever possible since it appears that these are ultimately the most significant ones in the context of Einstein's equations.

Index of main symbols

1
Introduction

A singularity, in the sense on which our later definitions will be based, is an exceptional or peculiar point in a space. For example, in global analysis a singularity in a smooth map from one manifold to another is (the image of) the places where the rank of the derivative is not a maximum – as contrasted with the "normal" situation of maximum rank. Or, to take a case closer to relativity, a singularity in a real-valued function which is everywhere else defined and continuous is an "exceptional" point at which the function cannot be given any value that makes it continuous throughout a neighbourhood of that point. As an example of this case, the electrostatic field of that hypothetical nineteenth century entity the "point charge"

$$\mathbf{E} = \text{const.} \times \mathbf{r}/|\mathbf{r}|^3$$

is singular, or "has a singularity", at $r = 0$.

In general relativity the term 'singularity' has undergone a succession of changes of meaning, which I shall sketch in historical sequence, introducing some of the basic definitions as we proceed.

1.1 The classical period

(I use the term 'classical' in its modern sense of 'before the author started doing research').

The first meaning follows the pattern of the case just described, of singularities in real-valued functions. If a metric is given in terms of components on a part of $\mathbb{R}^4$, then the singularities are the points of $\mathbb{R}^4$ at which one of the g_{ij} or g^{ij} cannot be continuously defined. One of the earliest known solutions to the vacuum Einstein equations contained singularities in this sense: the Schwarzschild metric in coordinates x, y, z, t on $\mathbb{R}^4$, having the form (after rewriting so as to display these Cartesian coordinates)

$$-\left(1 - \frac{2M}{r}\right) dt^2 + dx^2 + dy^2 + dz^2 + \frac{2M}{r(r - 2M)}(x dx + y dy + z dz)^2$$

(1)

(where $r = +\sqrt{x^2 + y^2 + z^2}$.)

This is singular on the 3-surface $r = 2M$ and on the 2-surface $r = 0$, in the sense that some components of g_{ij} cannot be defined there so as to give continuous functions.

Subsequently, singularities in this sense were found in the de Sitter metric

$$-dr^2 - R^2 \sin^2 (r/R) \left[d\theta^2 + \sin^2 \theta d\phi^2\right] + \cos^2 (r/R) c^2 dt^2 \quad (2)$$

as it was given in 1917, and in the Friedmann metric (discovered in 1922, but given here in its modern form):

$$-dt^2 + a(t)^2 \left[dr^2 + f(r) \left(d\theta^2 + \sin^2 \theta d\phi^2\right)\right] \tag{3}$$

where $f(r) = \sin^2 r$, r^2 or $\sinh^2 r$ and $a(t) \to 0$ as $t \to 0$. For both of these $\det(g_{ij})$ tends to zero on a 3-surface: for de Sitter, the surface $r = \pi R/2$, for Friedmann the surface $t = 0$. So on these surfaces some component of g^{ij} cannot be defined, giving a singularity in the sense at present under discussion.

From the start, however, there was dissatisfaction felt with the this notion of singularity, because it clearly depended on a particular choice of coordinates. Consequently, the assertion that a metric was singular, in this sense, might not correspond to anything physically measurable in the spacetime represented by the metric in question.

This was stressed by Einstein (1918) in his discussion of the de Sitter metric, where he pointed out that two conditions had to be fulfilled for a singularity to be real. First, the singularity had to be accessible, in the sense that there was a timelike curve leading from a regular point to the singularity and having a finite proper-time. Secondly, it must not be possible to find a new coordinate system with respect to which the metric becomes regular at the singularity and capable of being continued past it. These two conditions will be expanded in the next two sections and will form the basis for our definition of a singular space-time. The first condition was actually ill-expressed by Einstein when he required merely that the singularity be reachable in a finite proper time. For the finite time condition is actually no restriction at all: if it is possible to draw any timelike curve to the singularity, then, by wiggling the curve to make its speed close to the speed of light, it is possible to draw a curve of finite proper time. In the next section we shall see how to modify this condition so as to single out singularities at a "finite distance".

1.2 The idea of incompleteness

As a simple example consider the metric

$$- \left(1/t^2\right) dt^2 + dx^2 + dy^2 + dz^2 \tag{4}$$

which is singular (g_{ij} being undefined) on the plane $t = 0$ (in the $\mathbb{R}^4$ covered by the coordinates t, x, y, z). If an observer starting in the region $t > 0$ tries to reach the surface $t = 0$ by traveling, say, along the world-line $x = y = z = $ const. (which is clearly a geodesic), he will not reach $t = 0$ in any finite time – the surface is infinitely far into the future. Moreover, the fact that the singularity is not physically real can be seen by putting $t' = \log(-t)$ in $t < 0$ when the metric becomes

$$-dt'^2 + dx^2 + dy^2 + dz^2 \tag{5}$$

with $-\infty < t' < \infty$. In other words, the lower part of the space (and also the upper part) is just Minkowski space in disguise, and there is no singularity.

In his paper on the de Sitter metric just referred to, Einstein decided that the singularity was accessible (correctly), and that it was not possible to make the metric regular by a change of coordinates (incorrectly). He therefore deduced that there was a real singularity and, interestingly, he promptly rejected the solution as a consequence. The situation is the same with the singularity at $r = 2M$ in the Schwarzschild solution (1): it is accessible, but there is a change of coordinates for which it becomes regular.

In order to make Einstein's criterion for accessibility work, we can simply demand that there should be a timelike *geodesic* which reaches the singularity in a finite proper-time. Such a geodesic will have an endpoint on the singularity, in whatever coordinates are being used to describe the situation, but it will not have any endpoint in the regular part of the space-time. A space-time like this, containing a timelike geodesic which (when maximally extended) has no endpoint in the regular space-time and which has finite proper length, is called *timelike geodesically incomplete*. Clearly this property of incompleteness, which now has no reference in it to a particular coordinate system, is independent of what coordinates we use to describe the space-time.

It would certainly be convenient to be able to use arbitrary curves to decide whether or not a singularity is accessible. Indeed, this seems to be physically reasonable, because if any fairly well-behaved observer (i.e. having bounded acceleration) can reach the singularity in a finite proper time, then the singularity should still count as physically accessible, even if no geodesic observer can reach it. We can capture this idea mathematically by using a different parameter on curves, in place of proper-time, so as to achieve a definition that includes the world lines of observers with bounded acceleration. This new parameter is called the generalised affine parameter.

1.2.1 *Formalism*

In the next sections we shall develop some of the mathematical machinery for dealing with these ideas. The notation will broadly follow Hawking and Ellis (1973). Briefly, the space-time manifold is denoted by M, its metric by g (regarded as a bilinear function from pairs of vectors at the same point to the reals). Boldface letters will be used for arrays of any sort. We suppose that all our curves are described by maps from an interval into the space-time that are differentiable almost everywhere and rectifiable. Then we make the following formulation

Definition. The generalised affine parameter length of a curve $\gamma : [0, a) \to M$ with respect to a frame

$$\mathbf{E} = (E_a : a = 0, \ldots, 3)$$

at $\gamma(0)$ is given by

$$\ell_{\mathbf{E}}(\gamma) = \int_0^a \left(\sum_{i=0}^3 g\left(\dot{\gamma}, E_i(s) \right)^2 \right)^{1/2} ds \qquad (6)$$

where $\dot\gamma$ denotes the tangent vector $d\gamma/ds$ and $\mathbf{E}(s)$ is defined by parallel propagation along the curve, starting with an initial value $\mathbf{E}(0)$: that is, we impose

$$\nabla_{\dot\gamma}\mathbf{E}(s) \;=\; 0$$
$$\underset{a}{E}(0) \;=\; \underset{a}{E}$$

(We abbreviate this to g.a.p. length.)

Definition. A curve $\gamma : [0, a) \to M$ is *incomplete* if it has finite g.a.p. length with respect to some frame $\mathbf{E}$ at $\gamma(0)$. If $\ell_{\mathbf{E}}(\gamma) < \infty$, then if we take any other frame $\mathbf{E}'$ at $\gamma(0)$ we have that $\ell_{\mathbf{E}'}(\gamma) < \infty$. This is because the corresponding parallely propagated frames satisfy

$$\underset{i}{E'} = L_i{}^j \underset{j}{E}$$

for a constant Lorentz matrix L, and hence

$$\ell_{\mathbf{E}'} \le \|\mathbf{L}\|\ell_{\mathbf{E}}$$

where $\|\mathbf{L}\|$ denotes the mapping norm:

$$\|\mathbf{L}\| = \sup\left(\sum_j \left(L_i^j X^i\right)^2\right)^{1/2} \tag{7}$$

(with the supremum over all $\mathbf{X}$ with $|\mathbf{X}| = 1$ and $|\mathbf{X}|$ denotes the Euclidean norm of the components).

Definition. A curve $\gamma : [0, a) \to M$ is termed *inextendible* if there is no curve $\gamma' : [0, b) \to M$ with $b > a$ such that $\gamma'|[0, a) = \gamma$. This is equivalent to saying that there is no point p in M such that $\gamma(s) \to p$ as $s \to a$; i.e. that γ has no endpoint in M.

Definition. A space-time is *incomplete* if it contains an incomplete inextendible curve.

Discussion. We have now established definitions of geodesic incompleteness (which can be qualified by restricting to various sorts of geodesics) and incompleteness in the sense just defined. Clearly one can formulate many other possible definitions by restricting the sort of curve used in the definition of incompleteness. For example, a space-time is called timelike incomplete if it contains an incomplete timelike inextendible curve. The definitions of geodesic incompleteness and incompleteness are concordant because, since the components $g(\dot{\gamma}, \mathbf{E})$ of the tangent vector to a geodesic are constant, the affine length is proportional to the generalised affine parameter length. So a geodesic is incomplete with respect to its affine parameter if, and only if, it is incomplete in the sense defined above.

The Friedmann "big bang" models (3) are geodesically incomplete (and hence incomplete) because the curve defined by

$$\gamma (s)^0 = a - s$$

$$\gamma (s)^i = \text{constant} \qquad (i = 1, 2, 3) \tag{8}$$

is a geodesic which is incomplete, having no endpoint in M as $s \to a$. Minkowski space is not incomplete (a result which is not trivial (Schmidt 1973)). The region $r > 2M$ in the Schwarzshild metric (1) is incomplete, while the region $r > 0$ in (1) is not a space-time, since g is not defined at $r = 2M$.

Finally, we note that many writers use "singular" as synonymous with "incomplete"; although as we have seen incompleteness is only one of the criteria which must be fulfilled for there to be a true singularity. Incompleteness corresponds to Einstein's accessibility criterion for a singularity. We must now consider the other requirement, needed to rule out an apparent singularity ("singularity" in the sense we were considering in 1.2) arising merely from a bad choice of coordinates.

1.3 Extendibility

In 1924 Eddington showed that there was an isometry between the space-time M defined by the regions $r > 2m$ in the Schwarzschild metric (1) and part of a larger space-time M'. Incomplete curves in M on which $r \to 2m$ were mapped by this isometry into curves that were extensible in M': the singularity at $r = 2m$ was no longer present. So if we identify the Schwarzschild space-time with the part of the Eddington space-time M' with which it is isometric, we see that it is not just incomplete in the formal sense defined above: it actually had a piece missing from it, a piece that is restored in

M'. The singularity at $r = 2m$ is thus a mathematical artifact, a consequence of the fact that the procedure used to solve the field equations had fortuitously produced only a part of the complete space.

We note that, despite this, there are still some authors who regard the Schwarzschild "singularity" at $r = 2m$ as genuine; but this is only justified if (as done by Rosen (1974)) one uses a nonstandard physical theory in which there is some additional structure (such as a background metric) which itself becomes singular under the isometry of the metric into M', so that one structure, the metric or the background, is always singular at $r = 2m$.

The situation in Schwarzschild clearly contrasts with that of the Friedmann metrics (3). For these, on any of the incomplete curves (8) the Ricci scalar tends to infinity. For the smooth spacetimes that we are considering at the moment this is impossible on a curve which has an endpoint in the space-time, and so there can in this case be no isometric M' in which these curves have an endpoint. (Later we shall consider space-times in which the metric is not necessarily smooth, for which this does not hold.)

The singularity at $r = 2M$ in the Schwarzschild solution came to be called a "coordinate singularity", a term denoting any singularity in the sense of 1.2 which either did not give rise to incomplete curves, or which was such that incomplete curves tending to the singularity could be extended in some enlarged space-time. This larger space-time was constructed by applying a transformation to the coordinates specifying the original space-time, and extending the new coordinates (in modern terminology: by applying a diffeomorphism into a larger manifold). The Friedmann singularity, on the other hand, was termed a "physical" one, because a physically measurable quantity – the Ricci scalar – was unbounded on incomplete curves. On the whole I shall avoid the terms "coordinate" and "physical", since, while they convey important ideas, it is hard to give them precise definitions. Instead, I shall use the mathematical concept of extension to distinguish between the two types, the Schwarzschild space-time being extendible but the Friedmann one not so.

Definitions. An *extension* of a space-time (M, g) is an isometric embedding $\theta : M \rightarrow M'$, where (M', g') is a space-time and θ is onto a proper subset of M'.

A space-time is termed *extendible* if it has an extension.

The relation between extendibility and incompleteness is then expressed by the following result, showing that extendible space-times are timelike incomplete.

Proposition 1.3.1

If M has an extension $\theta : M \rightarrow M'$ then there is an incomplete timelike geodesic γ in M such that $\theta \circ \gamma$ is extendible.

Proof

Let $x \in \partial\theta M \subset M'$ and let N be a convex normal neighbourhood of x in M'.

Case 1. Suppose there exists a point y in $\theta M \cap (I^+(x) \cup I^-(x)) \cap N$. Then let γ' be the closed geodesic segment in N with $\gamma' : [0,1] \rightarrow N$, $\gamma'(0) = y$, $\gamma'(1) = x$. Let I be a connected component of the set $\{s | \gamma'(s) \in \theta M\}$ and let $\gamma = \theta^{-1} \circ \gamma' | I$. Then I is relatively open, non-empty and connected in $[0,1]$ and $1 \in I$. Hence either $I = (a,b)$ or $I = [0,b)$, and so $\theta \circ \gamma$ can be extended to b, as required.

Case 2. Suppose $M \cap (I^+(x) \cup I^-(x)) \cap N = \emptyset$. Then we can choose $y \in M \cap N \backslash (I^+(x) \cup I^-(x))$ and $x' \in (I^+(y) \cup I^-(y)) \cap (I^+(x) \cup I^-(x)) \cap N$. Thus $x' \in M$. Let γ' join y to x', with $\gamma'(0) = y$ and $\gamma'(1) = x'$. Then define I as before and continue as in Case 1.

□

1.4 The maximality assumption

The forgoing result has shown that if M is extendible then there is some timelike curve (actually a geodesic) – i.e. a possible worldline of a particle – which could continue in some extension of M but which in M itself simply stops. This seems unreasonable: why should M be cut short in this way? It seems natural to demand that "if a space-time can continue then it will"; in other words to demand that any reasonable space-time should be inextendible. This is an assumption imposed upon space-time in addition to the field equations of Einstein.

It can easily be shown that any space-time can in fact be extended until no further extension is possible. At this point the space-time is called maximal, and so we are led to the idea that we need only consider maximal space-times. But this idea is not really as innocuous as it might seem, because of the problem that an extension of a space-time, when it exists, cannot usually be determined uniquely. In special cases there are unique extensions: an analytic space-time has (subject to some conditions) a unique maximal analytic extension; similarly a globally hyperbolic solution of the field equations (with a specified level of differentiability) is contained in a unique maximal solution. In both these cases a sort of "principle of sufficient reason" demands that the maximal solution be taken. But suppose one has a non-analytic space-time where Einstein's field equations fail to predict a unique extension (either because there is a Cauchy horizon or because there is some sort of failure of the differentiability needed for the existence of unique solutions). Or suppose a situation arises in which there is a set of incomplete curves, each one of which can be extended in some extension of the space-time, but where there is no extension in which they can all be extended. (There exist, admittedly artificial, examples of this (Misner, 1967).) In cases such as these the same principle of sufficient reason would not allow one extension to exist at the expense of another. Perhaps the space-time, like Buridan's ass between two bales of hay, unable to decide which way to go, brings the whole of history to a halt.

Any solution to these problems can only come through a greater understanding of the physics of situations which might give rise to non-uniqueness. In the absence of strong enough physical theories to enable us to decide, we can only note here that maximality, while a useful instrumental principle that seems very likely, is far from being absolutely certain. Nonetheless, we shall usually adopt the principle.

1.5 Singularities

We are now in a position to give a definition of a singular space-time, that incorporates the ideas we have just described. This will supersede the more primitive idea of a singularity with which we started in 1.2.

Definition. A space-time is *singular* if it contains an incomplete curve $\gamma : [0, a) \to M$ such that there is no extension $\theta : M \to M'$ for which $\theta \circ \gamma$ is extendible.

According to this definition, the region $r > 2m$ in the Schwarzschild solution (1) is not singular, merely incomplete. To say that a space-time is singular means that there is some positive obstacle that prevents an incomplete curve continuing: it is not just that the space-time is smaller than it might be. Note also that we have not yet defined "a singularity", only "singular". This will be rectified in the next chapter, when it will appear that any singular space-time contains a singularity, and so in a maximal space-time *all* incomplete inextendible curves end at a singularity. For the time being, I shall occasionally refer to a "singularity", when speaking loosely, in anticipation of its definition later.

1.5.1 Singularity theorems

It turns out that most physically reasonable known exact solutions, when maximally extended, are singular, in the sense of the definition just given. Of those mentioned so far, only the de Sitter metric (the maximal analytic extension of (2)) is not singular, but this is an exceptional case among isotropic cosmologies.

When it was realised that most cosmological solutions were singular, reactions varied. It would appear that at first Einstein and Hermann Weyl took the view that a singularity in the metric could be interpreted as the presence of a singular matter-source and should be rejected on the grounds that one should only be interested in regular matter-sources; though later they turned this argument on its head and regarded particles of matter as being non other than singularities. Others felt that, though singularities were inevitable in any description of cosmology and astrophysics by means of very symmetrical metrics, they were an artifact of the high symmetry – on the analogy with Newtonian gravitational theory where a singularity (in the sense of 1.1) is obtained when a cloud of particles collapses from an initially spherical state, but there is no singularity when a general initial state is assumed. There would be little point in devoting one's energy to the study of singularities if the only singular space-times were unrealistically symmetric.

That this was not so was shown in work started by Geroch and carried out definitively by Hawking and Penrose, with important recent additions by Tipler. They showed that any space-time must be causally geodesically incomplete if it satisfied the following conditions (here only given very roughly: see (Hawking and Penrose, 1973) for full details).

1. Either

    (i) the null geodesics from some point are eventually focussed, or

    (ii) the null geodesics from some closed 2-surface are all converging.

2. M contains no closed timelike curves
3. For every non-spacelike vector K we have
$$R_{ab}K^a K^b > 0$$
4. Every non-spacelike geodesic, with tangent vector K, contains a point at which
$$K_{[a}R_{b]cd[e}K_{f]}K^c K^d \neq 0.$$

Whether or not (1) is satisfied is a contingent matter: there is ample evidence that (i) is satisfied for any past-directed null-cone in our universe, because of the focussing influence of the black-body radiation; while (ii) is more or less the definition of a black hole, whose occurrence somewhere in the universe is thought quite likely, and which Schoen and Yau (1983) have shown will occur whenever there is a large enough aggregation of matter. Condition (3) is satisfied for all plausible types of non-quantum matter (that is, excluding Dirac fields and such like), while (4) is a genericness condition serving only to rule out certain highly symmetrical solutions. (It is interesting that, contrary to what was at first thought, this means that the presence of symmetry makes it less likely, not more likely, that the space will be singular.)

We are left with condition (2), called the causality condition. (Space-times not satisfying (2) are said to exhibit *causality violation*). There is no direct evidence for the causality condition. However, cosmological arguments suggest that the part of the universe that we can see is close to a Friedmann model, which does satisfy condition (2), even though the universe as a whole may not do so.

But the Friedmann solutions contain a Cauchy surface (a spacelike surface intersected exactly once by all inextendible causal curves). So it is not unreasonable to suggest that our universe too, even though not exactly Friedmann, will contain some spacelike surface S without a boundary. Then the domain of dependence of S (the largest subspace in which S is a Cauchy surface) will satisfy the condition (2). This domain of dependence is denoted by $D(S)$, and it might be thought of as extending from S to the past back to near the big-bang singularity, and in the future up to near the occurrence of singularities caused by gravitational collapse. So the application of the theorem suggests that a part of our universe, $D(S)$, must be incomplete.

At this stage of the argument, one's conclusions depend on what attitude is taken to maximality (section 1.4). One possibility might be that $D(S)$ is a maximal Cauchy development (the largest space-time that can be determined by Einstein's equations on the basis of data on S) and that, even if the space-time is not singular, it comes to an end because there exists no determinate equation of evolution that can fix one continuation rather than another. On the other hand, if one accepts that space-time must be maximal then there are two other possibilities.

1. Space-time is not singular, but it violates condition (2). This causality violation could then bring $D(S)$ to an end, accounting for the failure of $D(S)$ to be the whole space-time. In this case, Tipler(1977) has shown that the causality violation must be of a very strange sort. Rather than a causality violating region emerging from a dense region where collapse has produced extreme physical conditions, as might be expected, the causality violating region starts "at infinity" in the sense of a generalised boundary point with an infinitely long curve ending at it. The problem in interpreting this is that it is entirely possible for an "infinity" in Tipler's sense to appear in the interior of a collapse, in which case it would be as bad as a singularity. If, however, we reject this then we are left with the second possibility,

2. Space-time is singular.

1.5.2 The nature of singularities

We have said in 1.5.1 that a space-time's being singular indicates some positive obstruction to the extension of a curve. The aim of this book is to provide clues as to the nature of this obstruction. What might "go wrong" at a singularity? There are two sorts of motivation for examining this question. First, we have just seen that any reasonable model representing our universe will contain a singularity, so that analysing singularities should tell us something about our universe. Second, if one has found a solution (either numerical or analytic) to Einstein's equations, and this solution turns out to be incomplete, then it is important to know whether there is an extension or whether the incompleteness is due to a true singularity (or, of course, there may be an extension in one region and a true singularity in another). To discover which is the case we need either a means for diagnosing singularities or, equivalently, a means for deciding when an extension exists. Ideally, it would be best to have a means for actually constructing extensions, when they exist. Thus, for both physical and mathematical reasons, we are led to investigate what it is that makes a spacetime singular; that is, what it is that prevents us from carrying out an extension.

The most familiar sorts of singularity are those in the Schwarzschild solution (1) as $r \to 0$ and the Friedmann solution (3) as $t \to 0$. In both these cases the scalar $R_{ijkl}R^{ijkl}$ constructed from the curvature becomes unbounded on curves ending at the singularity. Indeed, we have already noted (1.3) that the unboundedness of such a geometrically defined scalar is sufficient evidence for a true singularity, provided the space-time is smooth. So it would be tidy if some scalar were unbounded at every singularity: this would then immediately explain the incompleteness of curves running into the singularity and give us a simple means of distinguishing the incompleteness due to singularities from the incompleteness due to extendibility.

Unfortunately, this is not the case. Consider the Weyl solution (sometimes called the Curzon bipolar solution)

$$-e^{2\lambda}dt^2 + e^{2(\nu-\lambda)}\left(dr^2 + dz^2\right) + r^2 e^{-2\lambda}d\phi^2 \qquad (9)$$

where

$$\lambda = -m_1/R_1 - m_2/R_2$$

$$R_i^2 = r^2 + (z - z_i)^2$$

$$\nu = -m_1^2 r^2 / 2R_1^4 - m_2^2 r^2 / 2R_2^4 + \left(2m_1 m_2 / (z_1 - z_2)^2\right)$$
$$\left[\left(r^2 + (z - z_1)(z - z_2)\right) / R_1 R_2 - 1\right]$$

with m_1, m_2, z_1, z_2 constants (Synge, 1960, p. 314).

In this metric, as $r \to 0$ with $z_1 < z < z_2$ the values of all the scalars are bounded; but nonetheless it turns out that a real singularity is approached.

To prove this, the method is to take a small loop with $r = \varepsilon$, const, round $z = 0$, and choose a frame at the start of the loop. The generalised affine parameter length of the loop then tends to zero as $\varepsilon \to 0$. But if one calculates the Lorentz transformation relating the frame at the start to the frame after it has been parallely transported round the loop, then one finds that this tends to a constant matrix not the identity. But it is simple to show (cf. 2.2.1 below) that in a space-time without singularities, if the generalised affine parameter length of a loop tends to zero, then the Lorentz transformation it engenders tends to the identity. So the space-time cannot be extended so as to remove the singularity.

This example poses many problems, both mathematical and physical. Mathematically, it shows that the idea that singularities can be identified by the unboundedness of a scalar is inadequate. Physically, it requires us to interpret and understand the theoretical possibility of singularities in a neighbourhood of which physics is perfectly regular, with no unbounded densities or unbounded tidal forces. Singularities of this form have attracted interest as "cosmic strings" left over from the early universe (Vickers, 1986).

This book will develop some of the tools needed to understand this problem. The approach that I shall take will be based entirely on non-quantum general relativity. In part this can be justified on the principle that, whatever quantum theory of gravity emerges, it is likely that it will be of such a nature that classical solutions can shed light on corresponding quantum solutions; or at least, such that the parts of the classical solutions not immediately adjacent to the singularities will be physically relevant. But, in addition, the neglect of quantum effects can be justified in those cases with which we shall be mainly concerned in the latter part of the book, where we shall be examining regions where the curvature is not unbounded in order to decide whether there is a regular extension of the space-time. Once one has decided that a regular extension does not exist, then one can start to consider the possibility of quantum effects that may modify the nature of the singularity or in some sense remove it altogether.

2

The Riemann tensor

The Riemann tensor will play two related roles in our work: as the mathematical expression of tidal forces, and as the generator of Lorentz transformations. Both these stem from the equations for families of curves.

General Formalism. Throughout this chapter we shall be dealing with situations in which there is defined a field of pseudo-orthonormal frames, usually denoted by $(\underset{i}{E} : i = 0, \ldots, 3)$.

If X is a tangent vector, then by the norm of X, $|\mathbf{X}|$, we mean the Euclidean norm of the components of X in this frame (and similarly for covariant vectors). If Y is a linear map from, say, the tangent space to itself, then by the norm of Y, $\|Y\|$, we mean the mapping norm

$$\|Y\| = \sup_{|X|=1} |Y(X)|$$

while for a general tensor T we define the norm by expressions of the form

$$\|T\| = \sup_{|X|=|Y|=|Z|=1} |T_{ijk}X^iY^jZ^k|.$$

(Note that the two preceding definitions are consistent!)

The Riemann tensor first appears in differential geometry as a Lie-Algebra valued 2-form on the frame bundle which, if we have a specified frame-field, converts to a Lie-algebra-valued two-form on space-time whose components when acting on vectors X and Z are $R^i{}_j(X, Z) = R^i{}_{jkl}X^kY^l$. The norm of the Riemann tensor can be defined either as a mapping norm from pairs of vectors to elements of the Lie algebra, or using the above definition

16

on the components $R^i{}_{jkl}$. Later we shall define norms on skew pairs of vectors – bivectors – to make these consistent. There are places where occasional ambiguity can arise in the definition of the norms, which will be clarified at the time. Note that the result of taking a different norm is merely to alter the numerical factors in equations, whose values are usually immaterial.

2.1 Families of curves

Throughout this section indices will refer to components with respect to a pseudo-orthonormal tetrad, and so will be raised and lowered by the flat-space metric η.

Suppose $f : (x, y) \mapsto f(x, y)$ is a map of the rectangle $[0, A] \times [0, B]$ into M. We shall write $X := f_* \partial/\partial x$, $Y := f_* \partial/\partial y$. Choose a pseudo-orthonormal frame

$$(E, E, E, E)_{0 \ 1 \ 2 \ 3}$$

at the point $O = f(0, 0)$ and then propagate it over the image of f as follows: at points of the form $f(x, 0)$ we require

$$\nabla_X \underset{i}{E} = 0 \qquad (y = 0) \tag{1}$$

(thus parallely propagating **E** along the curve $x \mapsto f(x, 0)$), while elsewhere in the image of f we require

$$\nabla_Y \underset{i}{E} = 0 \tag{2}$$

(parallely propagating **E** along each curve $y \mapsto f(x, y)$, starting at $y = 0$).

Similarly, set $\mathbf{F}(O) = \mathbf{E}(O)$ and parallely propagate **F** first along $y \mapsto f(0, y)$ and then along $x \mapsto f(x, y)$ by setting

$$\nabla_Y \underset{i}{F} = 0 \qquad (x = 0) \tag{3}$$

$$\nabla_X \underset{i}{F} = 0 \tag{4}$$

Since $[\partial/\partial x, \partial/\partial y] = 0$ and Lie brackets are preserved by f_* we have

$$\nabla_X Y = \nabla_Y X \tag{5}$$

2.1.1 Geodesic deviation

Suppose now that

$$\nabla_Y Y = q^i(y)\, E_i,$$

with q^i independent of x. In other words, each of the curves $y \mapsto f(x, y)$, for various values of x, has the same history of acceleration in a parallely propagated frame. Set

$$X = X^i\, E_i.$$

Then

$$\nabla_Y \nabla_Y X \; = \; \frac{\partial^2 X^i}{\partial y^2}\, E_i$$

(from (2))

$$= \nabla_Y \nabla_X Y = \nabla_X \nabla_Y Y + R(Y, X)Y$$

(using (5) twice)

$$= q^i(y)\nabla_X E_i + R(Y, X)Y. \tag{6}$$

If we now write

$$\nabla_X E_i =: S_i{}^j\, E_j,$$

then we have

$$\nabla_Y \nabla_X E_i \; = \; \frac{\partial S_i{}^j}{\partial y}\, E_j$$

$$= R(Y, X)\, E_i = (\overset{\mathbf{E}}{R}{}^j{}_i (Y, X))\, E_j$$

(where $\overset{\mathbf{E}}{R}{}^j{}_i$ denote the elements of the curvature two-form defined on M by means of the tetrad $\mathbf{E}$). Consequently

$$S^j{}_i = \int_0^y \overset{\mathbf{E}}{R}{}^j_i(Y, X)dy'$$

so that (6) becomes

$$\frac{\partial^2 X^j}{\partial y^2} = q^i(y) \int_0^y \overset{\mathbf{E}}{R}{}^j_i(Y, X)dy' + \overset{\mathbf{E}}{R}{}^j_i(Y, X)Y^i \tag{7}$$

In the case where the curves $x = \text{const}$ are geodesics, we have

$$\nabla_Y Y = 0$$

and

$$q^i = 0,$$

so that (7) turns into the Jacobi equation:

$$\frac{\partial^2 X^j}{\partial y^2} = \overset{\text{E}}{R}{}^j{}_{ikl} Y^k X^l Y^i \tag{8}$$

expressing the acceleration of the vector X (giving the separation of "infinitessimally close" geodesics) in terms of the Riemann tensor. This acceleration is usually described as a *tidal force* tending to distort any material body that is following an approximately geodesic trajectory. The tidal force is, according to (8), liable to become unbounded if the components of the Riemann tensor are unbounded, and so special physical interest attaches to the boundedness or unboundedness of these components.

Equation (8) plays an essential role in the incompleteness theorems of Hawking and Penrose (Hawking and Ellis, 1973). From it one can derive equations for the rate of expansion and shear in a congruence of geodesics, and hence discover when geodesics contain pairs of conjugate points – a crucial step in proving that a space-time is incomplete.

Mathematically, the importance of this equation lies in the bounds which it enables us to put on the size of X in terms of the size of the Riemann tensor. We quote the following result whose proof is in (Clarke, 1982). A similar result (a version of the Sturm comparison theorem) was obtained by Tipler (1977).

Lemma 2.1.1

Suppose that the norm of the components of the Riemann tensor in equation (8) is less than some constant r_0 and that the initial values of X and $\dot{X}$ at $y = 0$ are X(0) and $\dot{X}(0)$. Write k for the norm of Y. Then:

1. For all y for which the equations are defined and the bound on R holds we have

$$|\dot{X}(0)| \left(2y - [\sinh{(\sqrt{r_0}yk)}/\sqrt{r_0}k]\right) + |X(0)| \left(2 - \cosh{(\sqrt{r_0}yk)}\right)$$

$$< |X(y)| < |\dot{X}(0)| \left[\sinh{(\sqrt{r_0}yk)}/\sqrt{r_0}k\right] + |X(0)| \cosh{(\sqrt{r_0}yk)}.$$

2. If in addition $X(0) = 0$ then for $0 < y < \pi/\sqrt{r_0}k$

$$|X(y)| > |\dot{X}(0)| \left[\sin{(\sqrt{r_0}yk)}/\sqrt{r_0}k\right].$$

Corollary 2.1.2

With the same bound on R and definition of k, if $X(0) = 0$ and $r_0 k < \pi/2$ then

$$|\dot{X}(1) - X(1)| < 1.32|X(1)|r_0 k^2.$$

Proof

Integrating (8) once gives

$$\dot{X}^j(s) - \dot{X}^j(0) = \int_0^s \overset{\mathbf{E}}{R}{}^j{}_{ikl} Y^k X^l Y^i dy. \tag{9}$$

So, using (a) of the lemma, we have that

$$|\dot{X}(s) - \dot{X}(0)| < \int_0^s \sqrt{r_0}k|\dot{X}(0)| \sinh(\sqrt{r_0}yk)dy$$

$$< (\pi/2)|X(1)|(\cosh(\sqrt{r_0}sk) - 1) \tag{10}$$

using (b) of the lemma to replace $|X(0)|$ by $|X(1)|$, together with the fact that $\sin x > 2x/\pi$ for $x < \pi/2$, and integrating.

Putting $s = 1$ and maximising $u^{-2}(\cosh u - 1)$ gives

$$|\dot{X}(1) - \dot{X}(0)| < |X(1)|r_0 k^2 \tag{11}.$$

Integrating the left hand side of (9) then gives

$$|X(1) - \dot{X}(0)| < \int_0^1 |\dot{X}(s) - \dot{X}(0)|ds$$

$$< \left(\frac{\pi}{2}\right)|X(1)|\left(\frac{\sinh(\sqrt{r_0}k)}{\sqrt{r_0}k} - 1\right)$$

from (10), which on maximising gives

$$|X(1) - \dot{X}(0)| < 0.32 r_0 k^2 |X(1)|.$$

Combining this with (11) then gives the result. $\qquad\square$

2.2 Holonomy aspects

Now define **L** to be the Lorentz matrix connecting **E** and **F**, by writing

$$\underset{i}{F} = L_i{}^j \underset{j}{E}$$

so that

$$\underset{j}{E} = L^k{}_j \underset{k}{F}.$$

Then using (2) we have

$$\nabla_X \nabla_Y \underset{i}{F} = \nabla_X \left(\frac{\partial L_i{}^j}{\partial y} \underset{j}{E} \right),$$

whence from (4) and (5)

$$R(X,Y)\underset{i}{F} = \frac{\partial^2 L_i{}^j}{\partial x \partial y} \underset{j}{E} + \frac{\partial L_i{}^j}{\partial y} \nabla_X \underset{j}{E}$$

$$= \frac{\partial^2 L_i{}^j}{\partial x \partial y} \underset{j}{E} + \frac{\partial L_i{}^j}{\partial y} \frac{\partial L^k{}_j}{\partial x} L_k{}^m \underset{m}{E}$$

which can be written

$$\frac{\partial^2 L_i{}^j}{\partial x \partial y} = \overset{\mathbf{E}}{R}{}^j{}_m(X,Y)L_i{}^m + \frac{\partial L_i{}^m}{\partial y} \frac{\partial L_k{}^j}{\partial x} L_m{}^k. \qquad (12)$$

Using (1) and (3), which imply that $L^j{}_i = \delta^j{}_i$ when either $x = 0$ or $y = 0$, this equation becomes, on integrating,

$$L_i{}^j(x,y) = \delta_i^j + \int_0^x \int_0^y \overset{\mathbf{E}}{R}{}^j{}_m(X,Y) L_i{}^m dx' dy'$$

$$+ \int_0^x \int_0^y \frac{\partial L_i{}^m}{\partial y} \frac{\partial L_k{}^j}{\partial x} L_m{}^k dx' dy'$$

the integrals being over the rectangle $0 \le x' \le x, 0 \le y' \le y$.

Since $\mathbf{E}(A,B)$ is defined by parallel propagation round two of the sides of the rectangle with $x = A$ and $y = B$, and $\mathbf{F}(A,B)$ by parallel propagation round the other two, the Lorentz transformation $\mathbf{L}(A,B)$ relating them is the transformation defined by parallel propagation round the entire perimeter. The above equation relates this to integrals over the area of the rectangle. If the rectangle is small then $L_i{}^j$ will be approximately $\delta^i{}_j$, while the product of first derivatives of $\mathbf{L}$ will be negligible, so that

$$L_i{}^j \approx \delta^j{}_i + \int \int R^j{}_i(X,Y) \, dx dy.$$

The next sections will make more precise this idea of the Riemann tensor generating Lorentz transformations. We shall first deal with the case of propagation round a "square", in the sense just discussed, and calculate the Lorentz transformation generated. Then we shall show that in suitable circumstances this enables us to generate arbitrary Lorentz transformations by propagation round loops of known size. (The estimate of the size is the crucial original element here). After this we derive an estimate for the effect of propagation round a "triangle", under less restrictive conditions on the Riemann tensor.

Lemma 2.2.1

Using the notation of the previous section, suppose that

$$|X(0)| = |Y(0)| = 1,$$

that the curves $x = $ const. are geodesics (so that (8) holds), and that ℓ is chosen so small that there exists an α, with $0 < \alpha < 1$, satisfying

$$\ell^2 \| \overset{\mathbf{E}}{\mathbf{R}}_0 \| = \alpha/28 \tag{13}$$

$$\ell \| \overset{\mathbf{E}}{\nabla \mathbf{R}} \| < \| \overset{\mathbf{E}}{\mathbf{R}}_0 \| / 20 \tag{14}$$

where $\overset{\mathbf{E}}{\mathbf{R}}_0$ and $\| \overset{\mathbf{E}}{\nabla \mathbf{R}} \|$, respectively, denote the value of the Riemann tensor at $x = y = 0$ in the frame $\mathbf{E}$, and the norm of the components of the covariant derivative in the frame $\mathbf{E}$. Equation (14) is required to hold throughout $0 < x < \ell$, $0 < y < \ell$. Then the Lorentz transformation Λ defined by parallel propagation round the square in the (x, y)-plane with side ℓ satisfies

$$\| \Lambda - \ell^2 \overset{\mathbf{E}}{\mathbf{R}}_0(X, Y) - \delta \| < \ell^2 \| \overset{\mathbf{E}}{\mathbf{R}}_0 \| / 5 < 6\alpha^2.$$

Proof

For simplicity we omit the label "$\mathbf{E}$" above Riemann components, since from now on all components are with respect to $\mathbf{E}$.

We first apply the lemma 2.1.2(b) with the initial conditions

$$X^j = X^j(0) \qquad \text{at } y = 0$$

and

$$\frac{dX^j}{dy} = \nabla_Y(\overset{j}{E} . X) = \overset{j}{E} . (\nabla_X Y) = 0 \qquad \text{at } y = 0$$

to give

$$|X| < |X(0)| \cosh(|Y| y \sqrt{r_0}) = \cosh(y \sqrt{r_0}) \tag{15}$$

whenever $\|\mathbf{R}\| < r_0$.

Moreover (13) and (14) together with $\|\mathbf{R}\| < \|\mathbf{R_0}\| + 2\ell\|\nabla\mathbf{R}\|$ give

$$\|\mathbf{R}\| < \alpha\ell^{-2}/25. \tag{16}$$

Now let ℓ' be the largest number not greater than ℓ such that in the square $0 < x < \ell'$, $0 < y < \ell'$ we have

$$\|\mathbf{L} - \delta\| < \alpha/15 \tag{17}$$

$$\left\|\frac{\partial\mathbf{L}}{\partial x}\right\| < \alpha^{1/2}\ell^{-1}/15, \qquad \left\|\frac{\partial\mathbf{L}}{\partial y}\right\| < \alpha^{1/2}\ell^{-1}/15 \tag{18}$$

where we know that such an $\ell' > 0$ exists because of (12) together with the boundary conditions

$$\mathbf{L}(0,0) = \delta, \quad \frac{\partial\mathbf{L}}{\partial x} = \mathbf{0} \text{ on } y = 0, \quad \frac{\partial\mathbf{L}}{\partial y} = \mathbf{0} \text{ on } x = 0. \tag{19}$$

We now show that in fact $\ell' = \ell$. For, if this were not so we should have equality in one of equations (17), (18) for some value of (x,y) in the range $0 < x < \ell'$, $0 < y < \ell'$. So to show that $\ell' = \ell$ it is sufficient to show that equations (16)–(19) imply strict inequality in (17)–(18).

First, (15) and (16) give

$$|X| < \cosh\left(\ell.\alpha^{1/2}\ell^{-1}/5\right) < 1.0201.$$

Inserting this together with (16)-(18) in (12) gives

$$\left\|\frac{\partial^2\mathbf{L}}{\partial x\partial y}\right\| < .048\alpha\ell^{-2}.$$

Whence integrating, using the boundary conditions (19), gives

$$\left\|\frac{\partial L}{\partial x}\right\|, \left\|\frac{\partial L}{\partial y}\right\| < 0.048\alpha\ell^{-1}, \qquad \|L - \delta\| < 0.048\alpha$$

which gives the required inequality in (17) and (18), showing that $\ell' = \ell$, i.e. that (17),(18) hold throughout $0 < x < \ell$, $0 < y < \ell$.

Now (12) can be written as:

$$\frac{\partial^2 L_i{}^j}{\partial x \partial y} - R^j{}_i(X,Y)\big|_0 = R_0{}^j{}_m(X_0,Y)(L_i{}^m - \delta_i{}^m)$$

$$+ \left(R^j{}_m(X,Y) - R_0{}^j{}_m + R_0{}^j{}_m(X - X_0, Y - Y_0) \right) L_i{}^m$$

$$+ \frac{\partial L_i{}^m}{\partial y} \frac{\partial L_k{}^j}{\partial x} L_m{}^k$$

where the subscript "$_0$" denotes evaluation at the origin. If we now integrate this over the rectangle $0 < x < \ell$, $0 < y < \ell$ and estimate all the terms by their bounds given in equations (16), (14) and (20), then we obtain the result. □

We shall now show that, under appropriate circumstances, any Lorentz transformation can be generated by parallel propagation round a loop, and obtain a specific estimate for the size of loop required. For the rest of this subsection we regard the components of the Riemann tensor in a given frame as constituting a mapping from the space of bivectors into the Lie algebra of the Lorentz group. If we place on the space of bivectors the norm

$$\|\mathbf{A}\|' = 2 \sup_{|X|=|Y|=1} |A_{ij} X^i Y^j|$$

and place the mapping norm on the Lie algebra, then the mapping norm for the Riemann components becomes

$$\|\mathbf{R}\| = \sup_{|X|=|Y|=|Z|=|W|=1} |R_{ijkl} X^i Y^j Z^k W^l|$$

in agreement with our earlier notation for the norm of $\mathbf{R}$. We shall suppose that the mapping defined by $\mathbf{R}$ is invertible, so that $\mathbf{R}^{-1}$ exists.

Let U be a compact (or precompact) region of the frame bundle and write

$$\varepsilon(U) = \inf_{E \in U} \min \left(\overset{E}{\|\mathbf{R}\|^2} \Big/ \left(1600 \overset{E}{\|\mathbf{R}^{-1}\|} \overset{E}{\|\nabla \mathbf{R}\|^2} \right), \right.$$

$$\left. 1 \Big/ \left(400000 \overset{E}{\|\mathbf{R}\|} \overset{E}{\|\mathbf{R}^{-1}\|} \right), 1/10 \right) \quad (21)$$

Then we have the following:

Lemma 2.2.2

Suppose $\mathbf{R}$ *is invertible, that* λ *is an element of the Lie algebra of the Lorentz group with* $\|\Lambda\| < 1/2$, *that* $\mathbf{E}$ *is a given frame and that*

$$U = \left\{ \mathbf{F} \ \middle| \ d(\mathbf{E}, \mathbf{F}) \overset{\mathbf{E}}{<} 32 \| \mathbf{R}^{-1} \|^{1/2} \| \lambda \| \right\}$$

(where d is the topological metric in the frame bundle - see 3.1.3 below). If $\|\lambda\| < \varepsilon(U)$, *then there exists a horizontal loop* κ *in* U *from* $\mathbf{E}$ *of length less than* $32 \| \mathbf{R}^{-1} \|^{1/2} \| \lambda \|$ *such that the Lorentz transformation* $\exp\lambda$ *is generated by parallel propagation round* κ.

Proof Set $\mathbf{L}' = \exp \lambda$. We number the conditions expressed by $\|\lambda\| < \varepsilon(U)$ and (21) as

$$\|\lambda\| < \|\mathbf{R}\|^2 / \left(1600 \| \mathbf{R}^{-1} \| \| \nabla \mathbf{R} \|^2 \right) \tag{22}$$

$$\|\lambda\| < 1/(400000\Gamma) \tag{23}$$

$$\|\lambda\| < 1/10 \tag{24}$$

where $\Gamma = \| \mathbf{R}^{-1} \| \| \mathbf{R} \|$ and all components are evaluated in the frame $\mathbf{E}$.

The bivector $\mathbf{R}^{-1}\lambda$ may not be simple, but there exist two independent dual simple bivectors related to $\mathbf{R}^{-1}\lambda$ by duality rotations, so that we can write

$$\mathbf{R}^{-1}\lambda = \mathbf{A} \cos \theta + *\mathbf{A} \sin \theta$$

and conversely

$$\mathbf{A} = \left(\mathbf{R}^{-1}\lambda \right) \cos \theta - * \left(\mathbf{R}^{-1}\lambda \right) \sin \theta.$$

Hence (since $\| * \mathbf{P} \|' < 2\sqrt{3} \| \mathbf{P} \|'$)

$$\|\mathbf{A}\|' < 4 \| \mathbf{R}^{-1}\lambda \|' < 4 \| \mathbf{R}^{-1} \| \| \lambda \| \tag{25}$$

and similarly

$$\| * \mathbf{A} \|' < 4 \| \mathbf{R}^{-1} \| \| \lambda \|.$$

We generate a first approximation to $\mathbf{L}'$ by parallel propagation round a square defined (as in the preceding section) by vectors X, Y with

$$|X| = |Y| = 1, \qquad \sum_i X^i Y^i = 0, \qquad \ell^2 X \wedge Y = \mathbf{A} \cos \theta$$

so that $\ell^2 = \|\mathbf{A}\|' \cos \theta$, i.e.

$$\ell < 2 \| \mathbf{R}^{-1} \|^{1/2} \| \lambda \|^{1/2}. \tag{26}$$

The constant α of equation (13) is thus

$$\alpha = 28\cos\theta\|\mathbf{R}(\mathbf{A})\| < 28\|\mathbf{R}\|\|\mathbf{A}\|' < 112\Gamma\|\lambda\| \qquad (27)$$

from (25). Also $\alpha < 1$ from (23), as is required.

Then from (26) we have

$$\ell\|\nabla\mathbf{R}\| < 2\|\mathbf{R}^{-1}\|^{1/2}\|\lambda\|^{1/2}\|\nabla\mathbf{R}\|$$
$$< \|\mathbf{R}\|/20$$

from (22), so that (14) is satisfied and lemma 2.2.1 can be applied.

Let Λ_1 be the Lorentz transformation achieved by parallel propagation round the square just defined, and Λ_2 the transformation achieved by propagation round the square similarly defined on replacing $\mathbf{A}\cos\theta$ by $*\mathbf{A}\sin\theta$.

Put

$$\mathbf{Z}_1 = \Lambda_1 - \mathbf{R}(\mathbf{A}\cos\theta) - 1$$

$$\mathbf{Z}_2 = \Lambda_2 - \mathbf{R}(*\mathbf{A}\sin\theta) - 1.$$

Then the lemma gives us that

$$\|\mathbf{Z}_i\| < 6\alpha^2 \qquad (i=1,2). \qquad (28)$$

The effect of parallely propagating round both squares, one after the other, will be expressed by the matrix $\Lambda = \Lambda_1\Lambda_2$ which satisfies

$$\Lambda - \mathbf{L}' = \mathbf{Z}_1(\mathbf{Z}_2 + 1 + \mathbf{R}(*\mathbf{A}\sin\theta)) + \mathbf{Z}_2(1 + \mathbf{R}(\mathbf{A}\cos\theta))$$
$$+ \mathbf{R}(\mathbf{A}\cos\theta)\mathbf{R}(*\mathbf{A}\sin\theta) - \sum_{r=2}^{\infty}\lambda^r/r!. \qquad (29)$$

Now

$$\|\mathbf{Z}_2 + 1 + \mathbf{R}(*\mathbf{A}\sin\theta)\| < 1 + 6\alpha^2 + 4\Gamma < 1.05$$

from (27) and (23), and

$$\|1 + \mathbf{R}(\mathbf{A}\cos\theta)\| < 1 + 4\Gamma\|\lambda\| < 1.05$$

again from (23). So from (29)

$$\|\Lambda - \mathbf{L}'\| < 6\alpha^2(2.1) + 16\Gamma^2\|\lambda\|^2 + \frac{\|\lambda\|^2}{2(1-\|\lambda\|)}$$

$$< \left(163088\Gamma^2 + 1\right)\|\lambda\|^2$$

(from (28) and (24))

$$< (17/40)\|\lambda\| + \|\lambda\|^2 < (21/40)\|\lambda\|$$

from (24).

Next we repeat the process, aiming to generate a transformation $\mathbf{L}''$ such that $\mathbf{\Lambda L}'' = \mathbf{L}'$, i.e. $\mathbf{L}'' = \mathbf{\Lambda}^{-1}\mathbf{L}'$. We verify that

$$\begin{aligned}
\|\mathbf{L}'' - 1\| &= \|\mathbf{\Lambda}^{-1}(\mathbf{L}' - \mathbf{\Lambda})\| < (21/40)\|\mathbf{\lambda}\|\|\mathbf{\Lambda}\| \\
&< (21/40)\|\mathbf{\lambda}\|(\|\mathbf{\Lambda} - \mathbf{L}'\| + \|\mathbf{L}'\|) \\
&< \big((21/40)\|\mathbf{\lambda}\| + \|\mathbf{\lambda}\| + \|\mathbf{\lambda}\|^2\big)(21/40)\|\mathbf{\lambda}\| \\
&< (18/160)\|\mathbf{\lambda}\|
\end{aligned}$$

from (24). Thus we can write $\mathbf{L}'' = \exp(\mathbf{\lambda}')$ with $\|\mathbf{\lambda}'\| < 1/2$, and

$$\|\mathbf{\lambda}'\| < 2\|\mathbf{L}'' - 1\| < (36/160)\|\mathbf{\lambda}\|. \tag{30}$$

Conditions (22)-(24) on λ are thus satisfied for $\mathbf{\lambda}'$, and the whole process can be repeated. We thus obtain a series of loops corresponding to a series $(\mathbf{\lambda}^{(n)})$ of Lie algebra elements, where $\mathbf{\lambda}^{(0)} = \mathbf{\lambda}$, $\mathbf{\lambda}^{(1)} = \mathbf{\lambda}'$, …generating, when applied one after the other, a sequence of Lorentz transformations that converges to $\mathbf{L}$. The combined length $\ell^{(0)}$ of the two loops generating $\mathbf{\Lambda}_1$ and $\mathbf{\Lambda}_2$ is given from (26) as

$$\ell^{(0)} < 4.4\|\mathbf{R}^{-1}\|^{1/2}\|\mathbf{\lambda}^{(0)}\|^{1/2}$$

and so the combined length of the whole sequence of loops is bounded by

$$16\|\mathbf{R}^{-1}\|^{1/2}\sum\|\mathbf{\lambda}^{(r)}\|^{1/2} < 32\|\mathbf{R}^{-1}\|^{1/2}\|\mathbf{\lambda}^{(0)}\|,$$

thus proving the lemma. $\qquad\square$

We can now state the following:

Theorem 2.2.3

Let $\mathbf{R}$ be invertible, let $U = \{\mathbf{F} \mid d(\mathbf{E}, \mathbf{F}) < \delta\}$, and let $\mathbf{L} = \exp(\lambda)$ be a given Lorentz transformation. Then there exists a curve in U of length less than

$$32\|\mathbf{R}^{-1}\|^{1/2}\|\mathbf{\lambda}\|^2\|\mathbf{L}\|^{11}\max\left(1/\varepsilon(U), 32\|\mathbf{R}^{-1}\|^{1/2}/\delta\right)$$

which generates $\mathbf{L}$.

Proof We generate the Lorentz transformation by generating n successive transformations $\exp(\lambda/n)$, where n is chosen so large that the previous lemma is applicable. This requires

$$\|\mathbf{\lambda}/n\| < \varepsilon(U)\|\mathbf{L}\|^{-11}$$

$$32\|\mathbf{R}^{-1}\|^{1/2}\|\lambda/n\| < \|\mathbf{L}\|^{-1}$$

where the factors of powers of $\mathbf{L}$ are to take care of the fact that each transformation starts from a different frame and so may have an altered value of $\mathbf{R}$ and of the other tensors involved in the lemma. In other words, we require

$$n > max\left(\|\lambda\|\|\mathbf{L}\|11/\varepsilon\,(U)\,,32\|\mathbf{R}^{-1}\|^{1/2}\|\lambda\|\|\mathbf{L}\|/\delta\right)$$

whence the result follows from the lemma.   □

2.2.1 Triangular case

We now deal with the case where, in the notation of 2.1.1, the separation vector (Jacobi vector) of the family of geodesics is zero at $y = 0$, so that the geodesics emanate from a single point. The argument is almost the same as that used in 2.2.3, but we reproduce it here for clarity.

Lemma 2.2.4 *Suppose that the conditions of section 2.1.2 hold; that $X(x,0) = 0$ and that $\sqrt{r_0}k < \pi/2$, $\sqrt{K} < 0.11$, where*

$$K = \max_{x}\left\{r_0\,|X(x,1)|\,k(x)\right\}.$$

Then

$$\left\|\frac{\partial\mathbf{L}}{\partial\mathbf{x}}\right\| < 4.61\sqrt{K}y^2 < 0.4, \qquad \left\|\frac{\partial\mathbf{L}}{\partial\mathbf{y}}\right\| < 9.21Kxy < 0.9.$$

Proof As before, we work in the region $0 < x,y < \ell'$, where ℓ' is the largest number less than 1 such that the following bounds hold

$$\|\mathbf{L}\| \quad < \quad 2 \tag{31}$$

$$\left\|\frac{\partial\mathbf{L}}{\partial\mathbf{x}}\right\| \quad < \quad Jy^2 + \varepsilon \tag{32}$$

$$\left\|\frac{\partial\mathbf{L}}{\partial\mathbf{y}}\right\| \quad < \quad 2Jxy + \varepsilon \tag{33}$$

$$J\ell'^2 \quad < \quad 1/8 \tag{34}$$

where

$$J = 2\sinh\left(\pi/2\right)K$$

and ε is any number with

$$\varepsilon < 1/8. \tag{35}$$

We show that equality cannot hold in (31),(32) or (33) under these conditions so that only (34) and (35) are in fact required. The result then follows on supposing that (34) is satisfied with $\ell' = 1$.

The results of 2.1.2 give immediately that

$$|\mathbf{X}(x,y)| < |\mathbf{X}(x,1)|y\sinh(\pi/2)$$

This, together with equations (31) – (34) allow us to write the equation (21) as the estimate

$$\left\|\frac{\partial^2 \mathbf{L}}{\partial \mathbf{x} \partial \mathbf{y}}\right\| < Jy + 2(Jy^2 + \varepsilon)(2Jxy + \varepsilon)$$

$$< Jy + 4J^2 xy^3 + \varepsilon$$

from (35).

If we integrate this with respect to x we obtain an estimate for $\partial \mathbf{L}/\partial y$ which can be written

$$\left\|\frac{\partial \mathbf{L}}{\partial \mathbf{y}}\right\| - 2Jxy - \varepsilon < Jxy\,(x/4 - 1) + \varepsilon x - \varepsilon$$

which implies that (33) cannot have strict equality. A similar argument holds on integrating with respect to y, whence the result follows as explained. $\qquad\square$

3

Boundary constructions

In the first chapter we defined a singular space-time as one containing incomplete inextendible curves that could not be continued in any extension of the space time. We must now give the definition (at times already anticipated) of the noun "singularity". The fundamental idea is that space-time itself (the structure (M, g)) consists entirely of regular points at which g is well behaved, while singularities belong to a set ∂M of additional points – "ideal points" – added onto M. We denote the combined set $M \cup \partial M$ by $Cl M$, the closure of M, and define the topology of this set to be such that phrases like "a continuous curve in M ending at a singularity p in ∂M", or "The limit of R as x tends to a singularity p is ..." all have meanings corresponding to one's intuitive picture of what they ought to mean.

The construction can be carried out in various ways and the set of ideal points, ∂M, could contain points other than singularities. Two important classes of ideal points that are not singularities are

1. endpoints of incomplete inextendible curves that can be continued in some extension of M (such endpoints being called regular boundary points) and

2. points "at infinity" such as I^+.

If the construction is carried out in such a way that $Cl M$ consists only of singularities and points of type (1) then ∂M will consist precisely of the endpoints of all incomplete curves. So the construction of such a set implies an equivalence relation on incomplete curves: namely, two incomplete curves can be defined as equivalent if they end at the same point in ∂M. Conversely, if we are given an equivalence relation $\approx$ between incomplete inextendible curves, then we can define ∂M to be the set

$$\{\text{incomplete inextendible curves}\}/ \approx$$

in such a way that the combined set ClM has a topology induced on it in a natural way.

Thus the problem of defining the boundary ∂M amounts to defining an equivalence relation $\approx$ on the set of incomplete inextendible curves, a relation that will become "has the same endpoint as" once ClM is constructed.

We note in passing that it would be possible to use any restricted class of curves, such as the set of incomplete inextendible timelike curves or geodesics, in the same way. Moreover, if we relax the condition on incompleteness and include some causal curves that are not incomplete in the construction, then we can obtain points at infinity.

Out of many proposals for defining the equivalence of curves we shall describe only two types, one due to Schmidt (1971) applicable to incomplete curves in general, and one due to Geroch, Kronheimer and Penrose (1972) applicable to causal curves (not necessarily incomplete).

3.1 The b-boundary

3.1.1 Horizontal lifts

Suppose $\gamma : [0, a) \to M$ is a smooth curve. If we parallely propagate a pseudo-orthonormal frame along γ, then the map that associates to each parameter value s the frame at $\gamma(s)$ is called a horizontal lift of γ. In other words, a map $\kappa : [0, a) \to LM$ is a horizontal lift of γ provided that $\pi \circ \kappa = \gamma$ and $D\kappa/Ds = 0$ (where π is the natural projection from LM to M, taking a frame at x into the point x).

If we fix a frame $\mathbf{E}$ to be $\kappa(0)$, then there is a unique horizontal lift of γ that starts at $\mathbf{E}$, in which case we can speak of the horizontal lift of γ through $\mathbf{E}$.

A *horizontal curve* $\kappa : [0, a) \to LM$ is any curve such that $D\kappa/Ds = 0$. Clearly a horizontal curve κ is the horizontal lift of the curve $\kappa_0 = \pi \circ \kappa$ through $\kappa(0)$.

We shall define the *b-length* of a horizontal curve

$$\kappa : [0, a) \ni s \mapsto \kappa(s) = (\underset{0}{E}, \underset{1}{E}, \underset{2}{E}, \underset{3}{E})$$

by

$$\ell(\kappa) = \int_0^a \left(\sum_{i=0}^3 g\left(\underset{i}{E}, \dot{\kappa}_0\right)^2 \right)^{1/2} ds.$$

We note that this is equivalent to the definition of g.a.p. length in 1.2.1, in that $\ell(\kappa)$ is simply $\ell_{\kappa(0)}(\kappa_0)$. Thus the horizontal lift of an incomplete curve will have finite b-length.

To formulate the next definition, recall that if

$$\mathbf{E} = \left(\underset{0}{E}, \underset{1}{E}, \underset{2}{E}, \underset{3}{E}\right)$$

and

$$\mathbf{E}' = \left(\underset{0}{E'}, \underset{1}{E'}, \underset{2}{E'}, \underset{3}{E'}\right)$$

are two frames at the same point, then there is a Lorentz matrix, which we denote by $\mathbf{L}(\mathbf{E}, \mathbf{E}')$, such that

$$L\left(\mathbf{E}, \mathbf{E}'\right)^i{}_j \underset{i}{E} = \underset{j}{E'}$$

3.1.2 b-equivalence

Suppose γ_1 and γ_2, each $[0, 1) \to M$, say, are two incomplete curves (not necessarily inextendible ones). We shall write $\gamma_1 \approx \gamma_2$ if there exist horizontal lifts κ_1, κ_2 and a sequence of horizontal curves λ_1, λ_2, ... such that

1. $\ell(\lambda_n) \to 0 \quad (n \to \infty)$
2. there are numbers s_n, t_n with $\lambda_n(0) = \kappa_1(s_n)$, $\pi\lambda_n(1) = \gamma_2(t_n)$
3. $\mathbf{L}(\lambda_n(1), \kappa_2(t_n)) \to 1 \quad (n \to \infty)$.

It is easy to see that $\approx$ is an equivalence relation.

The definition of equivalence just given is the form that will be used most often in this book. We shall see shortly that there is in fact a distance-relation on the frame bundle, and that points on equivalent curves do in fact get closer together, in the sense of this distance, as they approach the ends.

In the case where the curve γ_1 is extendible, with an end-point p in M, then the curves are equivalent precisely when the curve γ_2 is extendible and also ends at p. (Schmidt, 1971)

We can now use this equivalence relation to construct a boundary to M: or rather, we can construct directly the closure Cl_bM. To this end, let C be the set of incomplete curves in M and set $Cl_bM = C/ \approx$. Then there is a natural map from M into Cl_bM defined by taking each point p into the equivalence class of all curves ending at p. It can be shown (Schmidt, 1971) that this map is a homeomorphism into Cl_bM, and so we can identify M with its image in Cl_bM. In this case the boundary can be defined by $\partial M = Cl_bM \backslash M$, or $Cl_bM = \partial M \cup M$.

The important aspect of Cl_bM is its topology: it is this that expresses whether one is "near" to a singularity when the singularity is regarded as a point in the boundary ∂M. Using the construction we have just described, a topology can easily be defined via sequential convergence, in two stages. First, suppose that $s' = (s'_1, s'_2, \ldots)$ is a sequence of points in M (regarded as a subset of Cl_bM). Then for x in Cl_bM we write $s' \to x$ whenever there exists an incomplete curve $\gamma : [0, 1) \to M$ and a sequence $(t_1, t_2, \ldots)$ tending upwards to 1 in $[0, 1)$ such that $\gamma(t_n) = s'_n$ and $[\gamma]_{\approx} = x$. Secondly, if $s = (s_1, s_2, \ldots)$ is an arbitrary sequence in Cl_bM (i.e. not necessarily in M), then we set $s \to x$ if, for every subsequence s^* of s, there is a subsequence s' of s^* with s' in M and $s' \to x$.

This definition is chosen so that the relation $\to$ satisfies the axioms for sequential convergence (Birkhoff, 1937), and hence defines a topology on Cl_bM. Clearly every incomplete curve γ in M has the class $[\gamma]$ as its endpoint in Cl_bM, so that the points in ∂M really are the endpoints of incomplete inextendible curves.

We now note that the topology just defined induces the discrete topology on ∂M. For, any sequence tending to a point x in ∂M must eventually lie in $M \cup \{x\}$ (otherwise there would be a subsequence s^* contained in $M \backslash \{x\}$ which would not contain an s' tending to x); and so $M \cup \{x\}$ is open. Consequently $\{x\} = \partial M \cap (\{x\} \cup M)$ is open in the relative topology on ∂M, which is therefore discrete. We shall denote this topology on Cl_bM by $\mathcal{T}_d$.

In the next section we shall give a different construction in which it is natural to give Cl_bM a different topology, which is the one normally used. But before leaving the present approach via equivalence classes it is interesting to note one alternative topology that can be defined without difficulty. Recall that most "decent" topological spaces satisfy the following property, called regularity (Kelley, 1955):

R: For every point x and every neighbourhood U of x there is a closed neighbourhood V of x such that $V \subset U$.

It is easy to show that, for any space X with topology $\mathcal{T}$ there exists a unique topology $\mathcal{T}^R$ defined as the finest topology that is coarser than $\mathcal{T}$ and that satisfies condition **R:** indeed, $\mathcal{T}^R$ is simply the union of all regular topologies coarser than $\mathcal{T}$. In this way, starting with the space $(Cl_b M, \mathcal{T}_d)$, we can define a regular topology $\mathcal{T}_d{}^R$ on $Cl_b M$.

The usual topology, defined in the next section, is intermediate between $\mathcal{T}_d$ and $\mathcal{T}_d{}^R$, and in many cases it coincides with $\mathcal{T}_d{}^R$.

3.1.3   *The bundle construction*

In 3.1.1 we defined the length $\ell(\kappa)$ of a horizontal curve in the frame bundle. The idea of length suggests that of distance, suggesting in turn that we carry out the whole construction, including the assignment of a topology, by using lengths in the frame bundle LM. We can do this provided that we extend the measure ℓ of lengths of curves from horizontal curves in LM to all (rectifiable) curves. This will give the construction of $Cl_b M$ normally found in textbooks (e.g. Hawking and Ellis, 1973). What follows is a brief sketch of this construction , of which full details will be found in (Schmidt, 1971) or (Dodson, 1980).

There is a natural action of the Lorentz group on LM, whereby a Lorentz matrix **L** acts according to

$$\mathbf{E} \mapsto \left(L^i{}_j \underset{i}{E} \right)^3_{j=0} = \mathbf{L[E]}, \tag{1}$$

say. Thus for any element **l** of the Lie Algebra of the Lorentz group (i.e. any skew-symmetric matrix) and any frame **E** we can define a curve through **E** by

$$s \mapsto (\exp s\mathbf{l})[\mathbf{E}].$$

Since all points on this curve project to the same point in M, the tangent-vector to the curve at **E** is a vertical vector, $V(\mathbf{l})$, say. So by fixing a basis l_i for the Lie Algebra, we obtain a basis $V_i = V(l_i)$ of the vertical vectors at **E**.

The horizontal vectors at **E**, on the other hand, are mapped isomorphically onto the tangent space to M at the point p in question by the map π_* induced by the projection $\pi : LM \rightarrow M$, and so the vectors E_i of the frame itself, when mapped back into the horizontal subspace by the inverse of π_*, provide a basis for the horizontal vectors.

Now, given vectors X and Y in the tangent bundle to the frame bundle at a frame **E**, we can define an inner product between X and Y by

$$Q(X, Y) = \sum_{i=1}^{10} X^i Y^i$$

where the X^i and Y^i are the components of X and Y relative to the basis formed by the V_i and the $\pi_*^{-1} E_i$ together. Hence we can define the length of a tangent vector X to LM to be

$$|X|_b = (Q(X, X))^{1/2}.$$

Now given *any* curve $\kappa : [0, 1) \rightarrow LM$ we define the length of κ by

$$\ell(\kappa) = \int_0^1 |\dot{\kappa}(s)|_b ds$$

which clearly agrees with our previous definition of ℓ (3.1.1) when κ is horizontal (because then κ has zero components relative to the V_i). Using this we can define a distance-function between points **E** and **F** of LM by

$$d(\mathbf{E}, \mathbf{F}) = \inf \ell(\kappa)$$

where the infimum is over all curves κ connecting **E** and **F**.

With this distance function LM becomes a metric space. We can then take its Cauchy completion, $\overline{LM}$, defined by adding ideal limits for Cauchy sequences, a construction which gives $\overline{LM}$ a natural topology defined by the metric. Finally, it can be shown that the action of the Lorentz group (1) extends by continuity from LM to $\overline{LM}$, and so it is possible to define $Cl_b M$ by $\overline{LM}/\mathcal{L}$, with the quotient topology (the finest topology in which $\bar{\pi} : \overline{LM} \rightarrow$

Cl_bM is continuous). The open sets in Cl_bM are precisely the images under $\bar{\pi}$ of the open sets in $\overline{LM}$.

In order to show that the two constructions – equivalence classes of curves and Cauchy completion of LM – are in fact the same, one can set up a correspondence between them. Given an incomplete curve, one can take an infinite sequence of points on its horizontal lift not having an accumulation point, and this will define a Cauchy sequence in LM and hence a boundary point in $\overline{LM}$. Moreover, if two curves are equivalent according to the definition we have given (3.1.2), then it is not hard to show that the corresponding Cauchy sequences are equivalent, because the curves λ_n, together with vertical curves in the frame bundle connecting $\lambda_n(1)$ and $\gamma_2(t_n)$, constitute curves in LM whose length tends to zero, which is the condition for Cauchy sequences to have a common limit. Conversely, given a Cauchy sequence in LM, the points of the sequence can be joined up to give an incomplete curve in LM, though not necessarily a horizontal one. However, the definition of the metric is such that, if one applies a varying Lorentz transformation to the points on the curve so as to make it horizontal, then its length becomes less, and so it is still incomplete; also the horizontal curve can be so constructed as to end at the same point on the boundary of LM as the original curve. Thus each Cauchy sequence determines an incomplete horizontal curve, and hence a boundary point according to the first construction. Moreover, if two Cauchy sequences have the same limit, then reversing the argument of the previous paragraph shows that the corresponding horizontal curves are equivalent.

That, in the barest outline, indicates that the two approaches are the same.

3.2 Structure of the b-boundary

3.2.1 Fibre degeneracy

The space $\overline{LM}$ defined above (3.1.3) is in general not a fibre bundle because it is not locally the product of the base and a standard fibre: the "fibre" (as we shall nonetheless call it) $\bar{\pi}^{-1}(p)$ can vary from point to point on the b-boundary. The aim of this section is to study the nature of this fibre.

First recall that $\bar{\pi} : \overline{LM} \to \bar{M}$ is defined as the map taking a frame **E** in $\overline{LM}$ into the equivalence class $\{\Lambda\mathbf{E}|\Lambda \in \mathcal{L}\}$ under the action of the Lorentz group $\mathcal{L}$. Thus $\bar{\pi}^{-1}(p)$ consists of a set $\{\Lambda\mathbf{E}_0|\Lambda \in \mathcal{L}\}$ for some particular choice of $\mathbf{E}_0$ in $\overline{LM}$. So if we define $G(\mathbf{E}_0) = \{\Lambda \in \mathcal{L}|\Lambda\mathbf{E}_0 = \mathbf{E}_0\}$ then it is immediate that

1. $G(\mathbf{E}_0)$ is a group;
2. $G(M\mathbf{E}_0) = MG(\mathbf{E}_0)M^{-1}$, (i.e. altering the choice of $\mathbf{E}_0$ to $M\mathbf{E}_0$ only alters G to a conjugate subgroup)
3. the points of $\bar{\pi}^{-1}(p)$ are in 1-1 correspondence with the coset space $\mathcal{L}/G(\mathbf{E}_0)$.

The group $G(\mathbf{E}_0)$ is caled the *singular holonomy group* at $\mathbf{E}_0$. It can be shown to be closed (Clarke, 1978) so that the correspondence in (3) above is an isomorphism of manifolds with group action. If $G(\mathbf{E}_0)$ is not the trivial identity group, then we say that the fibre $\bar{\pi}^{-1}(p) = \mathcal{L}/G(\mathbf{E}_0)$ is *degenerate*.

Next we establish a condition under which $L \in G(\mathbf{E}_0)$. First a definition: Suppose $\kappa : [0,1] \to M$ is a loop (i.e. $\kappa(0) = \kappa(1)$) and that **F** is a frame at $\kappa(0)$. Let $\bar{\kappa}_F$ be the horizontal lift of κ defined by parallely propagating **F** round κ. Then $\bar{\kappa}_F(1)$ is related to **F** by the Lorentz transformation $L = L(\mathbf{F}, \bar{\kappa}_F(1))$. We shall write $h(\mathbf{F}, \kappa)$ for this transformation L so that $h(\mathbf{F}, \kappa)\mathbf{F} = \bar{\kappa}_\mathbf{F}(1)$.

Proposition 3.2.1 *Let a horizontal curve $\bar{\gamma} : [0,1) \to LM$ end at a point $\mathbf{E} \in \bar{\pi}^{-1}(p)$, with $p \in \partial_b M$. Then $L \in G(\mathbf{E})$ if and only if there exists a sequence $(t_i)_{i\in\mathbb{Z}} \to 1$ and loops $(\kappa_i)_{i\in\mathbb{Z}}$, $\kappa_i : [0,1] \to M$ satisfying*

$$\kappa_i(0) = \kappa_i(1) = \gamma(t_i) \qquad (2)$$

(where $\gamma = \pi \circ \bar{\gamma}$)

$$h(\bar{\gamma}(t_i), \kappa_i) \to L \qquad (3)$$

$$\ell_{\bar{\gamma}(t_i)}(\kappa_i) \to 0 \qquad (4)$$

Proof

1. Suppose that $L \in G(\mathbf{E})$. Then $L\bar{\gamma}$ and $\bar{\gamma}$ both end in $\bar{\pi}^{-1}(p)$ and thus they both end at the same point in $\overline{LM}$ (from the definition of $G(\mathbf{E})$ and of the action of the Lorentz group). From the definition of the Cauchy completion, this implies that there exist sequences (t_n), (t'_n) and $(\tilde{\kappa}_n)$ with $\tilde{\kappa}_n : [0,1] \to LM$, $\ell(\tilde{\kappa}_n) \to 0$, $t_n \to 1$, $t'_n \to 1$, $\tilde{\kappa}_n(0) = \bar{\gamma}(t_n)$, $\tilde{\kappa}_n(1) = L\bar{\gamma}(t'_n)$ (see 3.1.2). Let κ'_n be the horizontal lift of $\pi \circ \tilde{\kappa}_n$ through $\tilde{\kappa}_n(0)$, so that $\tilde{\kappa}_n(u) = \lambda_n(u)\kappa'_n(u)$ for some function $u \mapsto \lambda_n(u) \in \mathcal{L}$. Then the tangent vector $\dot{\tilde{\kappa}}_n$ will be given by

$$\dot{\tilde{\kappa}}_n = V\left(\dot{\lambda}_n\right) + \lambda_n(u)_* \dot{\kappa}'_n(u)$$

where $\dot{\lambda}_n$ is the tangent vector to the curve λ_n, regarded as an element of the Lie algebra of $\mathcal{L}$ and $V(\dot{\lambda}_n)$ denotes the corresponding vertical vector field on LM defined in 3.1.3. Consequently

$$\ell(\tilde{\kappa}_n) = \int \left(\|\dot{\lambda}_n\|^2 + \|\dot{\kappa}'_n\|^2\right)^{\frac{1}{2}} du$$

$$\leq \int \|\dot{\lambda}_n\| du + \int \|\dot{\kappa}'_n\| du.$$

Since $\ell(\tilde{\kappa}_n) \to 0$, this implies that $\int \|\dot{\lambda}\| du \to 0$ and that $\ell(\kappa'_n) = \int \|\dot{\kappa}_n\| du \to 0$. So $\|\lambda_n(1) - 1\| \to 0$. Now let $\bar{\kappa}_n$ be the horizontal curve defined by describing first κ', which runs from $\bar{\gamma}(t_n)$ to $\lambda_n(1)^{-1}L\bar{\gamma}(t'_n)$, followed by the curve

$$\kappa''_n : s \mapsto \lambda_n(1)^{-1}L\bar{\gamma}(t_{n'} + s[t_n - t'_n]) \qquad s \in [0,1].$$

Its projection $\kappa_n = \pi \circ \bar{\kappa}_n$ is a loop; we have that $h(\bar{\gamma}(t_i), \kappa_i) = \lambda_n(1)^{-1}L$, which tends to L; and

$$\ell(\kappa_n) = \ell(\kappa'_n) + \|\lambda_n(1)^{-1}L\|\ell(\bar{\gamma}|[t_n, t'_n]),$$

which tends to 0. Thus equation (2)—(4) are verified.

2. Now suppose, in order to prove the converse part, that for some L there exists a sequence (κ_i) satisfying equations (2)—(4). Write $u_i = \bar{\gamma}(t_i)$ and write $\bar{\kappa}_i$ for the horizontal lift of κ_i through u_i. Let $\mathbf{E}_0$ be the endpoint of $\bar{\gamma}$ and let $\mathbf{E}_1 = L\mathbf{E}_0$. Then we need to show that $\mathbf{E}_1 = \mathbf{E}_0$. Now, by definition of π, u_1 is the endpoint of $L\bar{\gamma}$. Choose n large enough so that (using (3)) there is a unique geodesic in $\mathcal{L}$ (regarded as a Riemannian

manifold) from $h(u_i, \kappa_i)$ to L for all $i \geq n$: call this geodesic ζ_i.
Then let σ_i be the curve in LM defined by describing:
first, $\bar{\kappa}_i$ from u_i to $h(u_i, \kappa_i)u_i$ then
second, the curve $s \mapsto \zeta_i(s)u_i$ from here to Lu_i. Then we have
that $\ell(\sigma_i) = \ell(\kappa_i) + \ell(\zeta_i) \to 0$; establishing that $L\bar{\gamma}$ and $\bar{\gamma}$ define
the same endpoint; i.e. that $\mathbf{E}_1 = \mathbf{E}_0$.

$\square$

3.2.2 Complete degeneracy

We see from the above proposition that if $G(\mathbf{E})$ is to be a proper
subgroup of the Lorentz group, then every Lorentz transformation
generated by loops of the form κ_i must (in the limit as the loops
approach the endpoint of the curve $\bar{\gamma}$) lie in the special subgroup
$G(\mathbf{E})$. And one would only expect this to happen if the space-
time had a special symmetry: the general case would be either
$G(\mathbf{E}) = \mathcal{L}$ or $G(\mathbf{E}) = \{e\}$. This expectation is confirmed by the
next proposition, which shows this to occur if the Riemann tensor
is surjective (regarded as a map from bivectors to the Lie algebra
of the Lorentz group) and if $\|\mathbf{R}^{-1}\| \to 0$ fast enough. We use the
notation of 2.2.

Proposition 3.2.2
*Suppose $\bar{\gamma} : (0, 1] \to \overline{LM}$ is a horizontal curve ending at $\mathbf{E}_0 \in$
$\bar{\pi}^{-1}(p)$ with $p \in \partial_b M$; and let there be given sequences of reals
$s_i \to 1$ and $\delta_i \to 0$. Let U_i be the open ball $\{\mathbf{F} \mid d(\mathbf{F}, \bar{\gamma}(s_i)) <$
$\delta_i\}$ and write $\mathbf{R}_i$ for the map defined by the components of the
Riemann tensor in the frame $\bar{\gamma}(s_i)$. If $\|\mathbf{R}_i^{-1}\|^{\frac{1}{2}} / \varepsilon(U_i) \to 0$ (with ε
as defined in 2.2) and $\|\mathbf{R}_i^{-1}\| / \delta_i \to 0$, then $G(\mathbf{E}) = \mathcal{L}$.*

Proof
Immediate from Prop. 3.2.1 and Theorem 2.2.3.

$\square$

For a regular point p of M one can define $E(p)$, the "emission"
of p, as the set of points that can be reached by null geodesics
from p. We can extend this definition to singular points, writing

$$E(p) = \{x \in M : \text{there is a null geodesic } \lambda_x \text{ from } x \text{ ending at } p\}$$

The next proposition shows that the neighbourhood-structure of $\partial_b M$ can be very strange.

Proposition 3.2.3

Let $p \in \partial_b M$ have a completely degenerate fibre. Then every neighbourhood of p contains $E(p)$.

Proof

Let $\mathbf{e}$ be the unique point of $\bar{\pi}^{-1}(p)$ and let U be any neighbourhood of p. By definition of the topology $\bar{\pi}^{-1}(U)$ is open and so there is a number ϵ such that the ball $B_\epsilon(\mathbf{e}) = \{x : d(x, \mathbf{e}) < \epsilon\}$ is contained in $\bar{\pi}^{-1}(U)$. Now consider some $x \in E(p)$. We can find a frame $\mathbf{F}$ such that, if we parametrise λ_x in such a way that the tangent vector has unit norm in $\mathbf{F}$, then λ_x reaches p at a parameter value less than ϵ. The horizontal lift, $\bar{\lambda}_x$ of λ_x through $\mathbf{F}$ terminates at $\mathbf{e}$ and has length equal to the parameter length, and so not greater than ϵ, and hence $\mathbf{F} \in B_\epsilon(\mathbf{e}) \subset \bar{\pi}^{-1}(U)$. Hence $x \in U$. $\square$

3.2.3 Robertson-Walker metrics

These provide much of the motivation for studying singularities, and so it is of particular interest to analyse the structure of their boundaries. As topological spaces they are the direct product of an open interval of $\mathbb{R}$ with a three dimensional manifold Σ, and the metric with respect to a coordinate t on $\mathbb{R}$ and local coordinates x^α on Σ takes the form

$$-dt^2 + a(t)^2 h_{\alpha\beta} dx^\alpha \otimes dx^\beta \qquad (\alpha, \beta = 1, 2, 3) \qquad (5)$$

where h is a metric on Σ such that (Σ, h) is a homogeneous space. The curves $x^\alpha = $ const. are the world-lines of fundamental observers, and such coordinates are described as "comoving" with the observers. For the "big bang" cosmological models, $a(t) \rightarrow 0$ as $t \rightarrow 0$ and the metric (taken as defined on, say $(0, b) \times \Sigma$) cannot be continued past $t = 0$. The curves $x^\alpha = $ const. are geodesics with t as affine parameter and so they terminate in a singularity to the past as $t \rightarrow 0$.

For consistency with our previous notation in which the singularity occurred at a value of the parameter equal to 1 we can reparametrise these curves by writing them as

$$\gamma_{\mathbf{x}} : s \mapsto (1 - s, \mathbf{x}) \qquad s \in [0, 1).$$

If in addition $a(t) \to 0$ as $t \to t_1 > 0$, then there is a singularity to the future as well as the past, encountered at $s' = 1$ by the curves parametrised as

$$\gamma'_{\mathbf{x}} : s' \mapsto (s' - 1 + t_1, \mathbf{x}) \qquad s' \in [0, 1).$$

(assuming that $t_1 > 1$).

The identification between these curves has been studied by Johnson (1977) and Bosshard (1976) (see also Dodson, 1980). Here we take a more general approach, not relying on the symmetries of the metric so much. Three different aspects of the boundary will emerge: the degeneracy of the fibres above the singularities; the identification of the curves λ_x for different $\mathbf{x}$; and the identification of $\lambda_{\mathbf{x}}$ with $\lambda'_{\mathbf{x}}$.

3.2.4 LM-neighbourhood structure

Choose an orthonormal field of frames (E', E', E') on the Rieman-
$_{123}$
nian manifold (Σ, h) and construct from these a field of frames (E, E, E, E) on (M, g) by setting (in the above coordinates)
$_{0123}$

$$E^{\mu} = \delta^{\mu}_0, \qquad E^{\mu} = a(t)^{-1} E'^{\mu}.$$
$$_0 _a _a$$

A straight-forward calculation (see, for example, Beem and Ehrlich (1981), p.71) shows that the non-zero rotation coefficients of the metric (5) are given by

$$\nabla_{\underset{\alpha}{E}} \underset{\beta}{E} = a^{-1} \gamma^{\delta}{}_{\alpha\beta} \underset{\delta}{E} + \frac{\dot{a}}{a} \delta_{\alpha\beta} \underset{0}{E}$$

$$\nabla_{\underset{\beta}{E}} \underset{0}{E} = \frac{\dot{a}}{a} \underset{\beta}{E}$$

where $\gamma^{\delta}{}_{\alpha\beta}$ are the rotation coefficients of the connection on (Σ, h).

Since these equations specify the horizontal curves in the frame bundle it is now quite a simple matter to prove directly that the fibre is degenerate, without recourse to the Riemann tensor. But in order to illustrate the use of the general result Prop. 3.2.2 we shall indicate how this proposition can be used to show degeneracy.

To determine the bundle distance metric we consider a curve $\gamma(s)$ with horizontal lift $\bar{\gamma}(s)$, and we define ρ^i to be the components of the timelike vector of the frame $\bar{\gamma}(s)$ with respect to the frame-field **E**, writing

$$(\bar{\gamma}(s))_0 = \rho^i E(\gamma(s)).$$

Then the horizontality condition reads

$$\frac{d\rho^0}{ds} = -\sum_{\alpha=1}^{3} \rho^\alpha \gamma'^\alpha \frac{\dot{a}}{a}$$

where the numbers γ'^α are given by

$$d\gamma/ds = \gamma'^\alpha \underset{\alpha}{E}.$$

Since ρ^0 specifies the magnitude of the Lorentz transformation $L = L(\bar{\gamma}(s), \mathbf{E}(\gamma(s))$ we have from this that

$$\frac{d\|L\|}{ds} \leq \frac{\dot{a}}{a}\|L\|\|\gamma'\|.$$

If now we suppose the parameter s to be chosen so as to measure b-metric length, then $|\gamma'| \leq \|L\|$ and the last equation becomes

$$\frac{d\|L\|}{ds} \leq \frac{\dot{a}}{a}\|L\|^2.$$

If we now take a general curve $\kappa(s)$ parametrised by b-metric length in LM and decompose its tangent vector into horizontal and vertical parts, then on applying the argument of Prop. 3.2.1(1) we find that the vertical component modifies the above equation to give

$$\frac{d\|L\|}{ds} \leq \frac{\dot{a}}{a}\|L\|^2 + 1.$$

Thus there exists a constant K such that if the length of the curve is less than $\min(Ka/\dot{a}, 1)$ then we shall have $\|L\| < 2$. From this we deduce:

Proposition 3.2.4

Suppose k is a curve in LM parametrised by b-metric length, and let t_0 be the time coordinate of the initial point. Let σ_1 be a parameter value with $\sigma_1 < \min(t_0, 1, Ka/\dot{a})$. Then the time coordinate of $\kappa(\sigma_1)$ is greater than $t_0/2$.

3.2.5 Applications

The choice of a particular equation of state for the matter in a model of the universe will determine the function a. As an illustration we shall take $a(t) = t^\alpha$ for $2/3 < \alpha < 1$, which, though not corresponding to any particular equation of state likely to hold up to near the singularity, still illustrates the general sort of behaviour. In this case $\dot{a}/a = t/\alpha$ and so in the previous proposition when $t_0 < 1$ we can take $\sigma_1 < K't_0$ where $K' = \min(1/4, K/\alpha)$.

We now apply Proposition 3.2.2 using a sequence of points tending to the singularity at $t = 0$ with $\delta_i = K't_i$, where t_i is the t-coordinate of the i'th point in the sequence. By the previous proposition this will ensure that in each U_i, t is bounded below by $t_i/2$ and $\|L\|$ is bounded above by 2.

The components of the Riemann tensor in the frame $\mathbf{E}$ (and hence in U) are easily calculated and we find that

$$\|\mathbf{R}\| < N_1 t^{-2}$$
$$\|\mathbf{R}^{-1}\| < N_2 t^{2\alpha}$$
$$\|\nabla \mathbf{R}\| < N_3 t^{-3}$$

Consequently for small t we can take $\varepsilon(U_i) = N_4 t^{2-2\alpha}$. We then have

$$\|\mathbf{R}_i^{-1}\|^{\frac{1}{2}}/\varepsilon(U_i) < \text{const.} \times t^{3\alpha-2} \to 0 \qquad (\text{for } \alpha > 2/3)$$

and

$$\|\mathbf{R}_i^{-1}\|/\delta_i < \text{const.} \times t^{2\alpha-1} \to 0.$$

Hence by 3.2.2 the fibres are completely degenerate.

Although we have derived this using a proposition that requires, in effect, that the derivative of the Riemann tensor should not be too large (so that it is possible to calculate the effects of curvature without things getting out of control) we should expect that in a more realistic model in which R was more chaotic it would be even more likely that that the Lorentz transformations generated by the Riemann tensor would fill up all the Lorentz group. So fibre degeneracy is likely to be a characteristic of general cosmological models.

3.2.6 *Identifications of curves*

The identifications between $\gamma_{\mathbf{x}}$ and $\gamma_{\mathbf{y}}$ follow from combining the degeneracy of the fibre with particular properties of the Friedmann solutions. Taking again $a = t^{\alpha}$ for simplicity we choose coordinates so as to write the metric on Σ as

$$h_{\alpha\beta}dx^{\alpha} \otimes dx^{\beta} = dr^2 + f(r)\,d\omega^2$$

where $d\omega^2$ is the metric of the Euclidean unit 2-sphere.

A past-directed null-geodesic λ from the point $t = t_0$, $r = 0$ satisfies

$$\frac{dr}{ds} = \frac{t^{-2\alpha}}{\sqrt{2}t_0^{-\alpha}}, \qquad \frac{dt}{ds} = -\frac{t^{-\alpha}}{\sqrt{2}t_0^{-\alpha}}$$

for a parametrisation in which $|\dot{\lambda}| = 1$ in the frame $\mathbf{E}$ at $t = t_0$, $r = 0$ (call it $\mathbf{E}_0$). So it will reach the singularity at $t = 0$ at a parameter value $s_1 = \sqrt{2}t_0/(1 + \alpha)$; at which stage r will have reached $r_1 = t_0^{1-\alpha}/(1 - \alpha)$. If $\mathbf{x}$ denotes the limiting value of the spatial coordinates of λ as $s \to s_1$, then it is simple to show that λ terminates at the same boundary point as $\gamma_{\mathbf{x}}$. Let $\mathbf{e}$ be the unique element of the fibre over this point. Then since the affine parameter is equal, in our parametrisation, to the b-metric distance, the distance from $\mathbf{E}_0$ to $\mathbf{e}$ satifies

$$d(\mathbf{E}_0, \mathbf{e}) \le s_1 = \frac{\sqrt{2}\,[(1 - \alpha)\,r_1]^{1/(1-\alpha)}}{(1 + \alpha)}$$

So if $\mathbf{x}$ and $\mathbf{y}$ correspond to the points in Σ with polar coordinates $(r = r_1, \theta = 0)$ and $(r = r_1, \theta = \pi)$, respectively, then the distance between the respective points $\mathbf{e}_{\mathbf{x}}$ and $\mathbf{e}_{\mathbf{y}}$ is

$$d(\mathbf{e}_{\mathbf{x}}, \mathbf{e}_{\mathbf{y}}) \le 2s_1 = \text{const.} \times r_1^{1/(1-\alpha)}.$$

Because of homogeneity, this holds for any two points having the distance apart of $\mathbf{x}$ and $\mathbf{y}$ in (Σ, h), namely $2r_1$.

Now, given any two such points $\mathbf{x}$ and $\mathbf{y}$ we can interpolate $(n - 1)$ points $\mathbf{x}_1, \mathbf{x}_2, \ldots, \mathbf{x}_{n-1}$ on the geodesic joining $\mathbf{x}$ and $\mathbf{y}$ in (Σ, h), the distance of consecutive points being $2r_1/n$. Travelling via these interpolated points thus gives

$$d(\mathbf{e}_{\mathbf{x}}, \mathbf{e}_{\mathbf{y}}) \le \sum_{r=1}^{n} d(\mathbf{e}_{\mathbf{x}_{n-1}}, \mathbf{e}_{\mathbf{x}_n}) \le \text{const.} \times r_1^{1/(1-\alpha)} n^{[1 - 1/(1-\alpha)]}$$

(where $\mathbf{x}_0 \equiv \mathbf{x}, \mathbf{x}_n \equiv \mathbf{y}$). On taking the limit as $n \to \infty$ we obtain $d(\mathbf{e_x}, \mathbf{e_y}) = 0$. So any two past boundary points are identified and there is in fact only one past boundary point.

Proposition 3.2.5

For a Robertson-Walker model with $a = t^\alpha$, $2/3 < \alpha < 1$, every neighbourhood of the past singularity contains the whole of M.

(cf. Johnson, 1977, for a somewhat stronger result.)

Proof

Every point in any Robertson-Walker solution can be joined to the past singularity by a null geodesic. In the case considered there is only one past singularity, so the result follows by Prop. 3.2.3. □

3.2.7 Past and future identification

If the situation described in the foregoing results holds for both past and future singularities, so that they are single points with degenerate fibres, then since the past and future singularities can be joined by a null geodesic of arbitrarily short affine parameter length, a repetition of the argument of Prop. 3.2.3 shows that these singularities have zero distance and so are identical.

3.3 Causal boundaries

The b-boundary is well adapted to studying extensions, because when an extension is possible, then the b-boundary comes as close as anything can to describing the topological boundary that the space-time will acquire when it is extended. However, when an extension is not possible, so that there is a genuine singularity, then as we have seen in the previous section the b-boundary may give a very curious picture of the structure of the singular points: it may turn out that there is only one singular point to which all incomplete curves run; or that the only neighbourhood of a singular point is M itself. In these circumstances, it is often preferable to use a different construction based on causality.

We define two curves to be equivalent in a new sense, and write $\gamma_1 \approx^\pm \gamma_2$ whenever $I^\pm(\gamma_1) = I^\pm(\gamma_2)$, where $I^\pm(\gamma)$ denotes the set composed of all points that are to the future (respectively, past) of some point of γ. Then we apply one of these equivalences, say $\approx^-$, to the set of all future directed timelike curves, whether complete or incomplete, extendible or inextendible. If two such curves are extendible (i.e. they have endpoints in M) then it turns out when (M, g) is strongly causal that they are equivalent precisely when their endpoints are the same (Geroch, Kronheimer and Penrose, 1972). This then sets up a correspondence between equivalence classes of extendible timelike curves and points in M (their endpoints). If we identify a point in M with the equivalence class of curves ending at that point, then we can regard M as a subset of the set $\hat{M}$ of equivalence classes of all timelike curves. The boundary $\dot{M} = \hat{M} \setminus M$ is then the future causal boundary of M. The past causal boundary is defined similarly by using past directed curves and the relation $\approx^+$. All this does, however, depend on the assumption of strong causality.

The set $\hat{M}$ contains equivalence classes of complete curves and of incomplete curves. Indeed, some equivalence classes may contain both sorts. When Penrose used this causal construction in discussing cosmic censorship (see section 6.2.5 below) he considered incompleteness with respect to proper time, and not with respect to the generalised affine parameter as we have done here. To distinguish these we shall call them "duration" and "length", respectively. Since any causal curve can be approximated arbitrarily closely by a piecewise smooth curve consisting of segments that are nearly null, and so can have arbitrarily small proper duration, it is clear that any class p in $\hat{M}$ contains members of finite duration (proper-time-incomplete curves). On the other hand, it could be the case that p contains only curves of finite duration; in this case, Penrose calls p "singular". Now, the duration is less than the length, so if all the members of p have finite length, then they will have finite duration and so the point will be singular in this sense (call it P-singular). But the converse is not true: it can happen that p is P-singular, but it has no members of finite length. An example occurs when M is the region $\{(t, x, y, z) \mid t > f(x)\}$ in

Minkowski space, where

$$f(x) = B + \int_0^x \frac{(x'^4 + 2x'^2 + 2)^{\frac{1}{2}}}{(x'^2 + 1)} dx'$$

and B is a constant chosen so that $f(x) - x \to 0$ as $x \to \infty$. Note that the curve $t = f(x)$ with y, z constant has finite duration. Let U be the region $\{(t, x, y, z) \mid x > t > f(x)\}$. Then it can be shown that the set of curves

$$p = \{\gamma \mid \gamma \text{ a future-directed causal curve with } I^-(\gamma) = U \}$$

is non-empty, so defining a point of $\hat{M}$, but that all the members of p have finite duration and infinite length. p is thus P-singular, but not a singularity from the point of view developed so far.

To obtain a boundary that is more analogous to ∂M we can restrict attention to *incomplete causal* curves. The space of equivalence classes in this case will be denoted by $\hat{M}_i$, with boundary $\dot{M}_i^-$, whose points will be referred to as *singular boundary points*. They are thus to be distinguished from causal boundary points at infinity (the remainder of $\hat{M}$), and we reserve the term *singularity* (in the sense of the causal boundary) for those particular singular boundary points that cannot be removed from the boundary by making an extension of the space-time.

There is one further complication that may arise in this approach: one can have a situation where a singularity extends out to infinity, as occurs, for example, in the Reissner-Nordstrom metric for $e^2 > m^2$ (Hawking and Ellis, 1973, p 160), where in the conformal picture the singularity at $r = 0$ intersects future null infinity at a point p which is in the causal boundary but not in $\dot{M}_i^-$, not being the end-point of an incomplete curve. There is a sense in which such points could be included in an extended definition of singular boundary points. To this end it is helpful to use the form of the construction originally used by (Geroch, Kronheimer and Penrose, 1972). First note that, since we have defined the points of $\hat{M}$ to be equivalence classes of curves γ having the same sets for $I^-(\gamma)$, each point of $\hat{M}$ is characterised by this set, and so we can identify the points of $\hat{M}$ with subsets of M having the form $I^-(\gamma)$ for some incomplete causal curve γ. We shall write M^* for the collection of sets of the form $I^-(\gamma)$ for a causal curve γ, whether γ is incomplete or not: these are precisely the indecomposable past sets (IPs) of the construction of Geroch, Kronheimer and Penrose. In this terminology we have the following:

Definition. An indecomposable past set A is a nonempty subset of M such that:

1. $I^-(A) = A$.
2. We cannot write $A = A_1 \cup A_2$ with A_1 and A_2 *proper* subsets satisfying (1).

The equivalence with our construction of $\hat{M}$ is then given by the following:

Lemma 3.3.1 *(Geroch, Kronheimer and Penrose, 1972) A set A is an IP if, and only if, it is of the form $I^-(\gamma)$ for some causal curve γ (which can be taken to be timelike).*

Proof

See Hawking and Ellis (1973) proposition 6.8.1, noting that this proof does not depend on whether or not γ is inextendible. $\quad\square$

When γ is extendible, terminating at a point p in M, then $I^-(\gamma) = I^-(p)$ can, as we have just noted, be identified with p providing that (M, g) is strongly causal. (The set is then called a PIP, for proper IP.) The subset of M^* consisting of sets of this form will be denoted by M^0, and can be identified with M. Finally, the difference $M^* \backslash M$ corresponds to the set of boundary points M^-. The construction can, of course be repeated interchanging past and future, and we can also restrict attention to IPs that are the pasts of *incomplete* causal curves, denoting the resulting sets by a subscript $_i$.

The advantage of rephrasing things in terms of IPs is that we have a natural causal structure on M^*. Indeed, there are two order-relations on the set that prove useful. First, we can simply define $p \le q$ for $p \subset q$. Clearly $\le$ is a partial order, analogous to the causality relation $p \in J^-(q)$ in space-time. Second, we can write $p \ll q$ if either $p = q$ or there exists an $x \in M$ with $p \subset I^-(x) \subset q$. This is more analogous to the chronology relation $p \in I^-(q)$. Though we shall use the terms "causal" and "timelike" for these orders, we have to remember that they must be taken with a pinch of salt: the terms do not have the same connotations as they do in space-time proper. Then a *timelike chain* in M^* is defined to be a set on which $\ll$ is a total order, viz. , for every p, q either $p \ll q$ or q$\ll$ p, while a *causal chain* is defined similarly for the $\subset$-order, $\leq$.

In practice the difference between the two stems from the fact that sets with inclusion as the order form a lattice with unique least upper bounds (l.u.b.'s). With $\gg$ the (timelike) l.u.b. is not unique; or, more precisely, given a set T there can be any number (including zero) of points p with the property that: p is an upper bound of T and there does not exist a $q \ll p$ which is also an upper bound of T. In such a situation we shall refer to any such p as *a* timelike l.u.b. for T. Concerning the existence of l.u.b.'s we have the following result:

Lemma 3.3.2 *If S is a causal chain in M^*, then the set*

$$p = \bigcup_{s \in S} s \stackrel{\text{def}}{=} \sum S$$

is an IP, and p is the causal least upper bound of S. If in addition S is a timelike chain, then p is also a timelike l.u.b. of S.

Proof Suppose S is a causal chain. Then p is a union of past sets (sets satisfying (1) of the definition of an IP) and so is clearly a past set itself. Suppose we could decompose p into $A \cup B$ with A and B proper past subsets with $A \backslash B$ and $B \backslash A$ nonempty. For any $x \in A \backslash B$, we can choose $s \in S$ with $x \in s$; then $s \subset A$, because otherwise the sets $s \cap A$ and $s \cap B$ would constitute a decomposition of s by proper subsets contrary to the definition of an IP. Similarly for any $y \in B \backslash A$ we have $y \in t$ with $t \subset B$. But by the definition of a chain, we have either $s \subset t$ or $t \subset s$; suppose $s \subset t$. Then we should have $x \in B$, a contradiction, and similarly if $t \subset s$. Thus p is an IP, and is obviously the least upper bound. If in addition S is a timelike chain, then it is again clear that p, constructed in the same way, is a least upper timelike bound. $\square$

In view of this, we can define a way of adding to M_i^* just those points that are causal l.u.b's of singular causal chains: we define the completion of M_i^* as

$$\bar{M}_i = \{p \in M^* \mid (\exists S)(\ S \text{ is a causal chain in } M_i^* \backslash M$$
$$\text{and } p = \text{l.u.b. of } S)\}$$

(It might seem more parsimonious to add, say, all l.u.b.s of timelike chains, on the grounds that these added points are not really singularities and so should only grudgingly be admitted. The definition above is adopted here because it results in a natural framework for our later arguments.)

The boundary $\bar{M}_i \backslash M$ is an extended causal boundary, consisting of the singular boundary points of M together with those points at infinity that are limits of singular boundary points.

In the case of greatest lower bounds the situation is somewhat less satisfactory, since the intersection $\prod S$ of a timelike or causal chain S of IPs is not necessarily an IP. However, if we choose an increasing maximal causal chain of IPs contained in $\prod S$ (provided it is not empty) then we obtain a "g.l.b." p as their union, in the sense that p is less than all the members of S and there is no other IP greater than p with this property. But g.l.b.'s , whether timelike or causal, are not unique.

3.4 The definition of a singularity

We have just seen that there are many possible definitions of boundaries, depending on the choice of curves to be used and the choice of an equivalence relation. For each definition of a boundary there is a corresponding definition of "singularity". Explicitly, a singularity in a boundary ∂M is a point in ∂M which is the endpoint of an incomplete inextendible curve γ, where γ is such that there is no extension θ of M for which $\theta o \gamma$ is extendible. In the rest of this book the b-boundary construction will be assumed, unless otherwise specified, and in this case when γ ends at p, then $[\gamma] = p$. Where the causal boundary is used, then we have, as just described, the notion of an extended boundary, and here we shall reserve the term *extended singularity* for the l.u.b. of a chain of singularities.

3.5 The A-boundary

The particular property of the A-boundary that makes it useful for us is the fact that the *manifold* can be extended through the entire A-boundary (whether or not any geometrical structure such as a metric can be so extended). It is thus an excellent candidate for describing a boundary that is not a true singularity, where we need just such an extension. The construction will be used in one theorem later on.

3.5.1 General idea

Suppose one specifies a particular atlas $\mathcal{A}$ on a space-time M. Each element of $\mathcal{A}$ comprises an open set W_i and a coordinate map $p_i : W_i \to \mathbb{R}^4$. For the applications we have in mind the coordinates p_i will be like normal coordinates, the atlas $\mathcal{A}$ being arranged so that the transition functions are not too large – as expressed by the condition (3) in the next section. Then one way to form a boundary for M is to take the closures of all the sets $p_i(W_i)$ in $\mathbb{R}^4$ and then "glue" the resulting sets back together again. The glueing obviously has to be done by the transition functions $\phi_{ij} = p_j \circ p_i^{-1}$ that relate the different sets $p_i(W_i)$ to each other, and so the construction is possible precisely when these transition functions have unique extensions to the closures of the $p_i(W_i)$.

In practice, it is asking too much to require that the whole of every one of these closures can be glued back. Instead we must suppose that, for at least some members of the atlas, there is a piece U_i^* of the boundary to $p_i(W_i)$ to which the maps ϕ_{ij} have unique extensions for all j, allowing the glueing-back to take place. If we have already arranged for the transition functions to have well-behaved derivatives (condition (3) below) then the additional condition needed on U_i^* to achieve this uniqueness is a Lipschitz condition restricting the wigglyness of the boundary of U_i (condition (2) below). By taking a fairly strong form for this condition we can also ensure that when the construction has been carried out the result of glueing together all the pieces is a manifold-with-boundary (proposition 3.5.3 below).

3.5.2 Detailed construction

Since no particular geometrical structure is being used we can formulate the construction for an arbitrary n-dimensional manifold, although we are interested in the case $n = 4$.

The boundary construction does not directly use any geometrical structure on the manifold other than the atlas, though we shall in fact be concerned with the case where the atlas being used is itself derived from a geometrical structure, so that the boundary is in fact geometrically determined.

The given atlas $\mathcal{A}$ is a collection $\{(W_i, p_i) \mid i = 1, 2, \ldots\}$ where $p_i : W_i \to \mathbb{R}^n$. Put $p_i(W_i) = U_i$. Then, in order to have a non-empty A-boundary we suppose that for each i there is given a subset U_i^* of U_i (which is non-empty for at least some i), a vector field k_i defined on a neighbourhood of U_i^* in U_i, and a constant K_i so that the collection of U_i and U_i^* satisfies the following properties.

1. U_i is compact.
2. For all $x \in U_i^*$, $|k_i(x)| = 1$ and there exists a number $\delta(x) > 0$ and a function $f_x : \mathbb{R}^n \to \mathbb{R}$ such that

$$y \in B\left(\delta(x), x\right) \cap U_i$$
$$\Longleftrightarrow \; (y \in B\left(\delta(x), x\right) \text{ and } k_i.(y - x) < f_x\left(P_{k_i} y\right)) \quad (6)$$

where $B(\delta, x)$ is the ball $\{y \mid |x - y| < \delta\}$, $P_k y = y - (k.y)k$, and $|f_x(y) - f_x(y')| < K_i|y - y'|$.
3. For each i and all j with $W_i \cap W_j = \leq$ there exist numbers K_i', K_i'' such that

$$\|D\phi_{ij}\| < K_i' \text{ and } \|D\phi^{-1}{}_{ij}\| < K_i''$$

where

$$\phi_{ij} = p_i p_j^{-1} : p_j(W_i \cap W_j) \to U_i.$$

In order that the construction be independent of a particular choice of U_i^*, we demand that U^* be maximal with respect to the existence of k_i, K_i as above.

The important condition here is (2), a Lipschitz condition on the boundary. For future use we shall give an alternative geometrical form of the condition:

Proposition 3.5.1

Condition (2) is equivalent to the following:

(2′) *For all* $x \in U_i^*$, $|k_i(x)| = 1$ *and there exists numbers* $\delta(x)$, $\theta(x)$, *both greater than* 0, *such that, for all* z *in* $B(\delta(x), x) \cap \bar{U}_i$,

$$y = z - \alpha k - v \in B(\delta(x), x), |v|/\alpha < \theta(x) \implies y \in U_i.$$

Proof

Throughout the following we shall simply write k for k_i and K for K_i.

First assume (2). Take $\theta = 1/(2K + 2)$. Suppose that y is as in the statement of (b′). Since $z \in \bar{U}_i$ we can choose a $z' \in U_i$ so that $|z - z'| < \alpha/(3K + 3) = \eta$, say. Then

$$k.(y - x) = k.(z' - x) + k.(y - z) + k.(z - z')$$
$$= k.(z' - x) - k.(\alpha k + v) + k.(z - z').$$

So

$$k.(y - x) - f_x(P_k(y)) < k.(z' - x) - \alpha(1 - \theta) + \eta - [f_x(P_k(y))$$
$$- f_x(P_k(z'))] - f_x(P_k(z'))$$
$$< -\alpha(1 - \theta) + \eta + K|P_k(y) - P_k(z')|$$

from (6)

$$= -\alpha(1 - \theta) + \eta + K|P_k(z - z' - k - v)|$$

which is less than zero because $P_k(k) = 0$ and because of the choice of the constants θ and η. Thus $y \in U_i$, as required by (2′).

Conversely, suppose that (2′) holds. Define $f_x(y) = \sup\{\alpha | y + \alpha k \in U_i\} - k.(y - x)$.

Then we can verify that $f_x(P_k(y)) = f_x(y)$ and that f_x is Lipschitz, as required for (2). □

3.5.3 Identification of limits

Set $U_{ij} = p_j(W_i \cap W_j)$ when $W_i \cap W_j \neq \emptyset$, otherwise $\emptyset$, and define

$$V_{ij} = \{x \in U_j \cup U_j^* \mid \exists \delta'(x) > 0 \text{ such that}$$
$$B(\delta'(x), x) \cap U_j \subset U_{ij}\}.$$

The construction is based on the following:

Lemma 3.5.2

If $(x_n)_{n \in \mathbf{N}}$ and $(x'_n)_{n \in \mathbf{N}}$ are sequences with $x_n \in U_{ji}$, $x'_n \in U_{ji}$ and $x_n \to x$, $x'_n \to x$ for some $x \in V_{ji}$, then the sequences $\phi_{ji}(x_n)$ and $\phi_{ji}(x'_n)$ also tend to limits and the limits are equal.

Proof

We continue to write k for k_i and K for K_i . We set

$$\begin{aligned}
\delta'' &= \min(\delta(x), \delta'(x)) \\
\epsilon_n &= |x_n - x| \\
\epsilon'_n &= |x'_n - x| \\
\epsilon''_n &= \max(\epsilon_n, \epsilon'_n) \\
\beta &= \min(1/2K, 1/2)
\end{aligned} \tag{7}$$

For large enough n we can ensure that $\epsilon''_n/\delta' < \beta$. So, for such n, define

$$\gamma_n(\alpha) = \begin{cases} x_n(1 - \alpha) + 2\alpha(x - \epsilon''_n k/\beta) & 0 < \alpha < 1/2 \\ x'_n(2\alpha - 1) + 2(1 - \alpha)(x - \epsilon''_n k/\beta) & 1/2 < \alpha < 1. \end{cases}$$

Then for $\alpha < 1/2$ we can apply the Proposition, with z replaced by x_n, y replaced by $\gamma_n(\alpha)$, α replaced by $2\alpha\epsilon''_n/\beta$ and v replaced by $2\alpha(x - x_n)$ to conclude that $\gamma_n(\alpha) \in U_i$, and so $\phi_{ji}(\gamma_n(\alpha))$ is defined for $0 < \alpha < 1/2$. A similar argument deals with the case $\alpha > 1/2$, using x' in place of x and $1 - \alpha$ in place of α .

The path γ has length not greater than $2\epsilon''_n(1 + 1/\beta)$ (from the triangle inequality applied to the two triangles $(x_n, x, x - \epsilon''_n k/\beta)$ and $(x'_n, x, x - \epsilon''_n k/\beta)$) and so its image under ϕ_{ij} has length not greater than $2K'(1 + 1/\beta)\epsilon''_n$ from condition (2), which tends to 0 as n tends to infinity. Hence

$$|\phi_{ji}(x_n) - \phi_{ji}(x'_n)| \to 0.$$

Similarly (using x_m in place of x', with $m > n$) we have that

$$|\phi_{ji}(x_n) - \phi_{ji}(x_m)| \to 0.$$

Hence the lemma. □

This now enables us to define $\bar{\phi}_{ji}$ on V_{ij} by the condition

$$\bar{\phi}_{ji}(x) = \lim_{n \to \infty} \phi_{ji}(x_n) \text{ for any } (x_n) \to x \text{ with } x_n \in U_{ij}.$$

Let $V_i = \bigcup_j V_{ji}$, and let

$$\sum_i V_i = \{(x, i) \mid x \in V_i\}$$

denote the disjoint union of the V_i. Define on this set a relation $\sim$ by

$$(x, i) \sim (y, j) \iff \bar{\phi}_{ji}(x) = y.$$

We now proceed to show that $\sim$ is an equivalence relation.

If $(x, i) \sim (y, j)$ and also $(y, j) \sim (z, k)$, then if $x_n \to x$ we have that $\phi_{ji}(x_n) \to y$, and hence that $\phi_{kj}(\phi_{ji}(x_n)) \to z$. But $\phi_{kj} \circ \phi_{ji} = \phi_{ki}$, i.e. $\phi_{ki}(x_n) \to z$ which means that $(x, i) \sim (z, k)$. Thus $\sim$ is transitive.

Now suppose that $(x, i) \sim (y, j)$ and that $y_n \to y$ with $y_n \in U_{ji}$. Then, from the lemma, $\phi_{ij}(y_n)$ tends to a limit, x' say. Choose any sequence x_n in U_{ji} with $x_n \to x$ and set $z_n = \phi_{ji}(x_n)$. Then $z_n \to y$ (since $(x, i) \sim (y, j)$) and so $\phi_{ij}(z_n)$ also tends to x'. But $\phi_{ij}(z_n) = x_n \to x$, and so $x' = x$. Thus $(y, j) \sim (x, i)$ and so $\sim$ is symmetric.

The relation is manifestly reflexive and so is an equivalence relation on $\sum_i V_i$.

Define

$$X = \left(\sum_i V_i\right) \Big/ \sim$$

and let P be the quotient map. Also let P_i be the map $x \to [(x, i)]_\sim$. For $x \in M$, define $\iota(x) = [(p_i(x), i)]$ for any i with $x \in W_i$. This is independent of the choice of i because of the definition of $\sim$.

Then ι is injective: for, suppose that $\iota(x) = \iota(y)$; i.e. suppose that we had $\bar{\phi}_{ji}(p_i(x)) = p_j(y)$ for some i and j. So, if $x_n \to p_i(x)$ with $x_n \in U_{ji}$ then we have that $\phi_{ji}(x_n) \to p_j(y)$. But $P_i(x_n) = P_j(\phi_{ji}(x_n)) = z_n$, say, so that $z_n \to x$ and $z_n \to y$. Since M is Hausdorff, $x = y$, proving that ι is injective.

Now define

$$\partial X = X - \iota(M); \quad V_i' = U_i \cup \operatorname{int}\left(V_i \cap P^{-1}(\partial X)\right),$$

where the interior is taken with respect to the relative topology on U^*, and define

$$\bar{M} = P\left(\sum_i V_i'\right), \qquad \partial_A M = \bar{M} - \iota(M).$$

Then $\partial_A M$ is the required A-boundary.

We give V' the relative topology induced from $\mathbb{R}^n$ and induce on $\bar{M}$ the quotient topology under P: the finest topology for which P is continuous.

Proposition 3.5.3
$\bar{M}$ is a manifold-with-boundary.

Proof

Let $x = [(y, i)]_\sim$ with $y \in V_i$ be an arbitrary point of $\bar{M}$ and let O be an open set containing x. By the definition of the topology on $\bar{M}$, $P^{-1}(O) \cap V_i$ is open (in the relative topology on V_i') and so we can define a neighbourhood T of y in this set, of the form $T = Q \cap V_i'$ where Q is an open neighbourhood of y in $\mathbb{R}^n$.

If $y \in U_i$ we can choose $Q \subset U_i$, while if $y \in V_i' \backslash U_i$ we can choose Q so that $Q \cap \dot{U}_i \subset V_i' \backslash U_i$. In the first case T can be taken to be homeomorphic to an open ball in $\mathbb{R}^n$, in the second case to the intersection of an open ball with a closed half-plane.

Conversely, let T be of one of these two forms. We will have proved M to be a manifold with boundary if we can show that $P(T)$ is open. From the definition of the quotient topology, $P(T)$ will be open precisely when $P^{-1}(P(T))$ is open. Now if $w \in P^{-1}(P(T))$ then $(w, j) \sim (x, i)$ for some $x \in T$; i.e. $\bar{\phi}_{ij}(w) = x$.

If $x \in U_i$ then $w \in P^{-1}(\iota(M))$ and so, from the definition of V_i', $w \in U_j$ with $\phi_{ij}(w) = x$. Since ϕ_{ij} is a homeomorphism we can find a neighbourhood of w mapped into T by ϕ_{ij} and hence in $P^{-1}(P(T))$.

On the other hand, if $x \in V_i' \backslash U_i$ we can show, by a similar argument, that $w \in V_j' \backslash U_j$. In this case, if it happened that $w \in [P^{-1}(P(T))]^\circ$, then there would be a sequence (w_n) with $w_n \in U_j \backslash P^{-1}(P(T))$ and $w_n \to w$. But $\phi_{ij}(w_n) \to x$, from the lemma, and x has an open neighbourhood in T, which is a contradiction.　　$\square$

This proves the proposition. It is clear from the proof that ι is a homeomorphism of M into $\bar{M}$, and that M is homeomorphic to $\iota(M)$. Thus ∂M is in the fullest sense a boundary of M.

Proposition 3.5.4
$\bar{M}$ is Hausdorff.

Proof

Suppose that x and y are not Hausdorff-separated, with, say, $x = P_i(x')$, $y = P_j(y')$, $x' \in V_i'$, $y' \in V_j'$. Then by taking open sets tending to x and y we can find a sequence (x_n) with $x_n \in P_i(U_i) \cap P_j(U_j)$ and $x_n \to x$ and $x_n \to y$. Let $x_n = P_i(x_n') = P_j(x_n'')$. Then $x_n' = \phi_{ij}(x_n'')$ and so $\lim x_n'' = \bar{\phi}_{ij}(\lim x_n')$. Thus, from the definition of $\sim$, $x = y$. $\qquad\square$

3.5.4 Extension through the A-boundary

First we define a covering of a neighbourhood of ∂M in $\bar{M}$ that is finer than the covering by the sets V_i'. For each i and each x in $V_i' \cap U_i^*$ let B_x^i be an open ball with centre x and radius less than $\delta'(x)$ such that $\bar{B}_x^i \cap \dot{U}_i \subset V_i'$, and let $C_x^i = P_i(B_x^i \cap V_i')$, $Q = \bigcup_{i,x} C_x^i$. The set Q is a subspace of the second-countable Hausdorff manifold-with-boundary $\bar{M}$, and so is itself second-countable, Hausdorff and regular and hence paracompact. So we can find a locally finite refinement $\{D_\alpha | \alpha \in A\}$ of the cover $\{C_x^i\}$ (see Kelley, 1955), where A is an index-set for the cover and $D_\alpha \subset \bar{W}_{j(\alpha)}$ for some function $j : A \to \mathbb{N}$.

From (Kobayashi and Nomizu, 1963) Appendix 3, Theorem 1 (which, although stated simply for manifolds, applies without change for manifolds-with-boundary), there exists a partition of unity $\{\xi_\alpha | \alpha \in A\}$ subordinate to the cover $\{D_\alpha\}$.

Now define

$$k = \sum \xi_\alpha P^*_{j(\alpha)} k_{j(\alpha)}, \qquad k_i'' = P_i^{-1} k$$

From the property (2) of the atlas it is clear that, in a neighbourhood of the boundary, $P_{j(\alpha)} k_j . k > 0$, and so k'' is non-zero.

We can now show that the vector $k' = k''/|k''|$ satisfies the same conditions (2) that the k_i did. Indeed, this is immediate from the form (2'). For k' is a convex combination of the vectors $\phi_{ij*} k_j$, and so the cone of vectors of the form $z - \alpha k + v$ (as in (2')) is contained in the union of the cones of this form for each of the vectors $\phi_{ij*} k$, which lie in U_i. Since each D_α is contained in a coordinate neighbourhood W_i we can, for each α, perform a smooth coordinate transformation so that, in D_α, the vector field k is simply $\partial/\partial x^n$.

This means that the transition functions between coordinate patches D_α (coordinates x) and D_β (coordinates x') will take the form

$$x^i = g^i(x'^1, x'^2, \ldots, x'^{n-1}) \qquad i = 1, 2, \ldots n-1$$
$$x^n = q(x'^1, x'^2, \ldots, x'^n).$$

As the covering $\{D_\alpha\}$ is a locally finite covering of a second-countable space, it is countable. So we can assume that the index α ranges over the integers and we can perform the extension by finite induction over α. By step number β in the induction we shall have defined a manifold with boundary M_β, a neighbourhood of the boundary being covered by patches D'_α ($\alpha = 1, 2, \ldots, \beta$) and D_γ ($\gamma = \beta + 1, \beta + 2, \ldots$). Each D'_α is an extension of D_α, the transition functions are extensions of the previously defined transition functions, and they retain the above form. Moreover the vector field $\partial/\partial x^n = k$ remains transverse to the boundary of the manifold.

The next step in the induction is as follows. Suppose x^i denote coordinates in $D_{\beta+1}$ and that the boundary $\partial D_{\beta+1}$ (corresponding to U_i^*) is

$$x^n = t\left(x^1, \ldots, x^{n-1}\right),$$

where t is a Lipschitz function and the domain is locally given by $x^n < t$. $D_{\beta+1}$ is extended by defining a new t by

$$t'(x) = t(x) + \epsilon\,\mathrm{dist}(x, \mathrm{edge}(\partial D_{\beta+1}))$$

where $\mathrm{edge}(S) = \bar{S}\backslash S$. Then for small enough ϵ, t' will still be Lipschitz.

Next, for each γ such that $\phi_{\gamma\,\beta+1}(D_{\beta+1})$ is not empty we extend the transition functions g and q (defined above) over the extended domain $D'_{\beta+1}$ by using the Whitney extension theorem (Clarke, 1982). This is done for each $\gamma = 1, 2, \ldots$ up to β, and after each one any further transition functions implied by the relation $\phi_{\delta\gamma} \circ \phi_{\gamma\rho} = \phi_{\delta\rho}$ are also defined. Finally, the manifold $M_{\beta+1}$ is defined as $M_\beta \cup D'_{\beta+1}/\approx$, where the relation $\approx$ is the identification defined by the coordinate map between $D'_{\beta+1}$ and M_β. The Hausdorff property for $M_{\beta+1}$ follows inductively from that for M_β and from the above construction.

Formally, the final extension is defined as the direct limit of the sequence $(M_1, M_2, \ldots)$ under the maps defined by the relation $\approx$.

However, because the covering by the Ds is locally finite, every point has a neighbourhood that is contained in (the image of) a single one of the manifolds M_β. Thus the direct limit is a Hausdorff manifold, and we have constructed an extension through the whole of the A-boundary.

3.5.5 Extension of geometrical structures

If we are given any geometrical structure defined by functions in the charts of $\mathcal{A}$, then if we extend through the A-boundary, the geometrical structures can usually be extended as well. In particular, suppose we are given a metric whose components are C^k in $\mathcal{A}$, with the derivatives having well-defined limits on the sets U_i^*. Then when each of the sets D_α is extended, the Whitney theorem allows us to extend the components of the metric over D'_α: for, if $p : D_\alpha \to M_\alpha$ is the map defined by the coordinates, then $p^*(M_\alpha)$ has a Lipschitz boundary in D_α, and so it is possible to extend the metric components in D_α, and then use p to map these back into $M_{\alpha+1}$. In this way the metric can be extended over the whole of the A-boundary.

4
Existence theory and differentiability

4.1 Introduction

Although many of our considerations will be purely geometrical, treating space-time as a pseudo-Riemannian manifold and asking whether or not this geometrical structure is breaking down, it must always be remembered that we are really working with a physical theory, governed by particular physical equations for fields and particles, and that it is the breakdown of the physics that is primarily of interest. The breakdown of the geometry is simply one possible manifestation of the breakdown of the physics.

Unfortunately there is a conflict between the mathematical contexts appropriate to, on the one hand, geometry and, on the other hand, physically significant differential equations. In differential geometry one deals with geodesics, domains of dependence and so on. For this to be valid one requires that the connection should satisfy a Lipshitz condition, which ensures the existence of unique geodesics and normal coordinate neighbourhoods. Providing this holds, the differentiability of the metric has little geometrical significance and it is customary to require it to be C^∞ for convenience. By contrast, in the study of hyperbolic differential equations (a type to which Einstein's equations belong) questions of differentiability are crucial. The differentiability chosen reflects the character of the solutions allowed: by choosing a low differentiability one admits solutions like shock-waves or impulse-waves which may be very significant; conversely, by choosing too high a level of differentiability one will brand as "singular" shock-wave solutions that from the point of view of fluid dynamics may be entirely legitimate.

If one accepts that our primary context is that of physical differential equations, then this should determine our basic choice of differentiability, and assumptions such as C^∞ would be regarded as idealisations convenient for geometrical purposes but of restricted validity. If our aim is to study genuine, physical singularities then we need to determine the circumstances in which further solution of some particular equations (Einstein-fluid, Einstein-Maxwell ...) becomes impossible. As we shall see below, this cannot yet be determined in general. But there are existence theorems which specify when a solution *is* possible: i.e. we know what is not a singularity, even though we do not yet know what is a singularity.

One consequence of these existence theorems is that there can exist solutions evolving from data in which the Riemann tensor components are unbounded in a compact set. So the unboundedness of the Riemann tensor is not sufficient for the existence of a singularity. If we work at the level of differentiability appropriate to these existence theorems then we have to accept unbounded Riemann tensors, and with them the non-uniqueness of geodesics and the non-existence of normal coordinate neighbourhoods.

In the next sections we shall review the different classes of differentiability conditions appropriate to geometry and to the existence theory of differential equations, before going on to review this theory itself. The aim will be to give the reader some feeling for the role played by different differentiability conditions, not to give any sort of complete account of the theory.

4.2 Differentiability

If we are given a space-time – that is, a differentiable manifold M, with a particular C^∞ atlas of charts specified and a metric whose components can be calculated in that atlas – then it is usual to classify the space-time according to the class of functions to which the components of the metric in any chart belong. For example, if these components are C^k then one speaks of a C^k space-time.

From some points of view, the Riemann tensor is of more direct physical importance than either the connection or the metric, in the sense that an unbounded Riemann tensor would suggest unbounded tidal forces, while an unbounded connection or metric component could be a purely coordinate effect. And so a physical distinction can be drawn between "well-behaved" space-times in which the Riemann tensor is bounded, and "badly behaved" ones in which it is not, even though, as we noted in the previous section, an unbounded Riemann tensor does not in itself indicate a singularity. However, the boundedness of the Riemann tensor does not imply that the metric is, say, C^{2-} (see below for definitions), and so this suggests that we should specify directly the boundedness of the Riemann tensor.

There is also a good mathematical reason for concentrating on the Riemann tensor to give a criterion for whether a space-time is well-behaved near a singularity. For each individual component of this tensor is an ordinary function on the frame-bundle LM, which is a space with a natural topological metric (see 3.1.3). So an invariant meaning can immediately be given to statements about whether the tensor is locally bounded, or Hölder continuous, etc. By contrast it is not usually possible to say that the metric has such properties near a singularity, because such statements can only refer to the components of the metric in a particular chart, and there may be no one chart that contains a neighbourhood of the singularity. So in much of our work "the differentiability of space-time" will refer to the differentiability of the Riemann tensor, rather than the metric.

4.2.1 C^k-classes

Recall that a function $f : \mathbb{R} \to \mathbb{R}$ is said to be Hölder continuous with exponent α if for each compact domain U there is a constant K with

$$|f(x) - f(y)| < K|x - y|^{\alpha}$$

for all x, y in U. If this holds with $\alpha = 1$ the function is called Lipschitz; we shall reserve the term Hölder-continuous for the case $0 < \alpha < 1$. The usual notation for differentiability classes is then as follows.

C^{∞}	functions having derivatives of all orders
C^k	functions having continuous derivatives of order k
C^0	continuous functions
$C^{k,\alpha}$	functions with a k'th derivative which is Hölder continuous with exponent α
C^{k-}	functions whose $(k-1)$st derivative is Lipschitz
C^{0-}	functions that are locally integrable and locally bounded.

When we want to indicate the domain ω on which the functions are defined then we write $C^k(\omega)$; otherwise the functions will be understood to be defined on all of $\mathbb{R}^n$ for some n. One appends a subscript $_0$ to the class to indicate that it is restricted to functions of compact support. Thus C_0^k functions with compact support having derivatives of all orders up to k.

From now on the index X will represent one of ∞, k, $k-$, k,α and $X \pm 1$ will represent, respectively, ∞, $k \pm 1$, $k\pm1-$, $k\pm1,\alpha$. By a "space-time" we refer to a triple $(M, \mathcal{A}, g)$ where $\mathcal{A}$ is an atlas on the Hausdorff, paracompact manifold M that is at least C^1 and g is the metric.

A space-time is said to be of class C^X if there is a C^{X+1} atlas $\mathcal{A}'$, C^1-equivalent to $\mathcal{A}$, in which the components of the metric are C^X. If a space-time is C^{2-} then there exist unique geodesics with a given initial vector, and so this is often taken as the minimum differentiability for any sort of geometry. As we shall see shortly, however, this is not the weakest condition under which geometry can be done.

As explained above, we shall emphasise the differentiability of the Riemann tensor. But since the differentiability of the metric cannot be deduced from that of the Riemann tensor, we need to place some minimal restrictions on the metric in addition to the Riemann tensor. We accordingly define a space-time to be C^X if there is an atlas C^1-equivalent to $\mathcal{A}$ in every chart of which

1. the metric components have C^{0-} weak derivatives
2. the Riemann tensor components are C^X.

(Recall that a weak derivative is a function that is a derivative in the sense of distributions.)

Then it is shown in (Clarke, 1982) that in a $\mathbf{C}^{0-}$ space-time there are unique timelike geodesics between nearby causally related points. This is probably the lowest differentiability class in which it is possible to perform conventional pointwise differential geometry.

In order to extend space-time through singularities that appear to be inessential, it transpires that one cannot stay within this class, however: if one uses any sort of C^k class then one must demand that space-time be $\mathbf{C}^{0,\alpha}$ – i.e. the Riemann tensor is not merely bounded, but Hölder-continuous. This has the disadvantage that any point where the Riemann tensor is unbounded, or even discontinuous, is regarded as a true singularity, thus excluding many types of discontinuity (such as the boundary of a fluid) that are perfectly acceptable physically. In some circumstances one can avoid this by working with "piece-wise" conditions, but this turns out to be impractical in the study of singularities.

We now turn to the differentiability classes appropriate to the solution of Einstein's equations, and in particular Sobolev classes.

4.2.2 Sobolev classes

The central result here is the theorem of Hughes, Kato and Marsden (1977) that if one uses harmonic coordinates to write Einstein's equations as a system of evolution equations for the components of the metric – so that the basic variables become the functions g_{ij} and dg_{ij}/dt, each functions on $\mathbb{R}^3$ and evolving in time – then existence can be proved when these functions on an initial $\mathbb{R}^3$-hypersurface lie in the Sobolev spaces $H^{s+1}(\mathbb{R}^3)$ and $H^s(\mathbb{R}^3)$ respectively, for some $s > 1.5$. (More accurately, one has to choose a background metric that specifies the boundary conditions at spatial infinity and then require that the difference between the true metric and this background metric belong to a Sobolev class.) The space $H^s(\omega)$, where ω is an open set, is defined by completing the space $C_0^{m+1}(\omega)$ (where m is the largest integer $\leq s$) with respect to a norm $\|f\|_{s,\omega}$. When s is an integer the norm is defined as

$$\|f\|_s = \left\{ \sum_{|\mathbf{i}| \leq m} \int_\omega \left(D^{\mathbf{i}} f \right)^2 dx \right\}^{\frac{1}{2}}.$$

Here $\mathbf{i}$ is a *multi-index*; i.e. a sequence $(i_1, i_2, \ldots, i_n)$ of nonnegative integers and we define

$$|\mathbf{i}| = \sum i_k, \qquad D^{\mathbf{i}} f = \frac{\partial^{|\mathbf{i}|} f}{\partial x_1^{i_1} \ldots \partial x_n^{i_n}}$$

If s is not an integer then

$$\|f\|_{s,\omega} = \left\{ \sum_{|\mathbf{i}| \leq m} \int_\omega \left(D^{\mathbf{i}} f \right)^2 dx \right\}^{\frac{1}{2}}$$

$$+ \left\{ \sum_{|\mathbf{i}|=m} \iint_{\omega \times \omega} \left| D^{\mathbf{i}} f(x) - D^{\mathbf{i}} f(y) \right|^2 / |x - y|^{n+2(s-m)} dx dy \right\}^{\frac{1}{2}}$$

$$(1)$$

We note that the form of this is close to the definition of the Hölder classes, so that much of the same techniques can be used. In the case where ω is $\mathbb{R}^n$ this norm is equivalent to one that can be expressed (Stein, 1970) in terms of the Fourier transform $\hat{f}$ of f as

$$\|f\|_s = \left\{ \int_{\mathbb{R}^n} |\hat{f}(k)|^2 \left(1 + |k|^2 \right)^s d^n k \right\}^{\frac{1}{2}} \qquad (2)$$

This form is obviously identical (apart from multiplicative constants) to the expression already given in the case of integer s, and suggests that there is a sense in which the norm for non-integral s interpolates between the integral values.

This idea of interpolation can be made more precise and more generally applicable. Roughly speaking, one wants to show that if something is true for H^n and for H^{n+1}, then it is also true for each H^s with $n < s < n+1$. This holds provided that the "something" can be expressed in terms of the norm of a linear map. In the general situation, suppose that we have two Banach spaces B_1 and B_2 (e.g. H^n and H^{n+1}) both contained in some common space X (in this example, $X = B_1$) and such that $B_1 \cap B_2$ is dense in each of B_1 and B_2. Suppose that we have a linear map L taking $B_1 \cap B_2$ into a space $\tilde{B}_1 \cap \tilde{B}_2$ where $\tilde{B}_1$, $\tilde{B}_2$ and $\tilde{X}$ have the same properties as B_1, B_2 and X, respectively, and suppose that we have estimated the size of the image of L for both B_1 and B_2, in the form

$$\|Lv\|_{\tilde{B}_1} \leq K_1 \|v\|_{B_1}$$
$$\|Lv\|_{\tilde{B}_2} \leq K_2 \|v\|_{B_2}$$

Then we can interpolate between B_1 and B_2 on one side, and between $\tilde{B}_1$ and $\tilde{B}_2$ on the other side, to obtain an interpolated estimate on the norms: given any θ between 0 and 1 there exist spaces T and $\tilde{T}$ in $B_1 + B_2$ and $\tilde{B}_1 + \tilde{B}_2$ with norms in between those on B_1 and B_2, such that L extends uniquely to T and satisfies

$$\|Lv\|_{\tilde{T}} \leq K_1^{1-\theta} K_2^{\theta} \|v\|_T$$

(For the proof see (Adams, 1975)).

When applied to H^n spaces these interpolating spaces are precisely the H^s. Moreover, in this case, because the H^n are Hilbert spaces, it is possible (see Palais (1968)) to give a much simpler description of the interpolation procedure than is the case for general Banach spaces.

Suppose that $B_1 \supset B_2$ and that both are Hilbert spaces, and denote the corresponding norms by $\| \ \|_1$ and $\| \ \|_2$. Then

$$v \to (\|v\|_1)^2 \qquad (v \in B_2)$$

is a bounded positive quadratic form on B_2 and defines a bounded positive self adjoint operator A by $(\|v\|_1)^2 = (v, Av)_2$. To construct the interpolation space, we define an intermediate norm on B_2 by $(v, A^s v)_2$ and then complete B_2 in this norm. The above interpolation theorem is then fairly straightforward to prove.

Finally, we note an alternative form of the norm (2), which we can write as

$$\|f\|_t = \left\{ \int_{\mathbf{R}^n} |f_1(k)|^2 \, d^n k \right\}^{\frac{1}{2}}$$

where

$$f_1(k) := \hat{f}(k) \left(1 + |k|^2\right)^{t/2}$$
$$= \int d^n x e^{ikx} \left(1 - \Delta^2\right)^{\lfloor t/2+1 \rfloor} (G_\alpha * f)(x)$$

and where the Bessel potential $G_\alpha(x)$ is defined by

$$\hat{G}_\alpha(k) = \left(1 + |k|^2\right)^{-\alpha/2}$$

(Stein, 1970) with $\alpha/2 = t/2 - \lfloor t/2 - 1 \rfloor$ and $\lfloor x \rfloor$ denoting the largest integer less than x; so that

$$G_\alpha(x) = \frac{2^{1-n/2-\alpha/2}}{\pi^{n/2} \Gamma(\alpha/2) |x|^{(n-\alpha)/2}} K_{(n-\alpha)/2}(|x|)$$

K_ν being a modified Bessel function.

4.3 Why Sobolev spaces?

What determines the choice of a particular function space in proving an existence theorem for Einstein's equations? The choice of Sobolev spaces rather than spaces of C^k functions is bound up with the use of the energy-inequality for a system of hyperbolic equations. Suppose we were to use the space of C^k functions. This is determined by the norm

$$\|f\| = \max_k \sup_x |D^k f(x)|,$$

involving finding the supremum of the value of a derivative. But, given a solution of a hyperbolic system, there is no simple relation between the maximum size of the derivatives of the data at points on an initial surface, and the maximum size of the derivatives of the solution evaluated at points in the domain of dependence. But there is a relation between the size of an integral involving the derivatives of the data and a similar integral for the solution. (This integral is referred to as an energy, because in certain special cases it does have this physical interpretation). The size of the energy integral is then directly related to the norm in a Sobolev space.

After deciding to use a Sobolev space, one must fix the order of the derivatives to be used, i.e. the index s of H^s. The decisive factor in determining the order of this derivative lies in the non-linearity of the equations, where there appear products of the metric, its inverse (the contravariant metric) and its derivatives. Given that the metric lies in one function space, we need to control the space in which its inverse lies and the space in which these products lie. The lowest differentiability class that can be used is fixed by the requirement that the product of two first derivatives of the metric (such as occurs in the Riemann tensor through the product of two Christoffel symbols) should be of no worse differentiability than a single first derivative. If this is achieved then the differentiability level of the terms in Einstein's equations follows the same pattern as obtains in a linear equation. We shall see in the next section that this forces the differentiability order of the first derivative to be greater than 1.5; i.e. the metric has to lie in a function space with differentiability order greater than 2.5. We note that, since we are dealing with Sobolev spaces, not C^k spaces, this does not mean that the metric can be differentiated even twice.

4.3.1 Products

We now outline the proof of this basic result concerning products of functions in Sobolev spaces. We shall consider functions defined on $\omega \subset \mathbb{R}^n$; the extension to a general manifold is easily done by using theorems that enable one to restrict functions to smaller domains and extend them to larger domains. (Adams, 1975)

Theorem 4.3.1

Let $f \in H^t$ and $g \in H^u$ (with the domain ω understood throughout). Let fg denote the product function $(fg)(x) = f(x)g(x)$. If $s < t + u - n/2$, $s \leq t$ and $s \leq u$ then $fg \in H^s$.

Proof We work first with C^∞ functions of compact support contained in ω for which the Fourier transform version of the Sobolev norm is applicable. We show that the norm of fg in H^s can be estimated by the norms of f and g in H^t and H^u respectively. The result then follows by completion of the set of C^∞ functions in these norms. Thus we need to estimate the integral

$$A = \|fg\|_s = \int \left| \left(\widehat{fg} \right)(k) \right|^2 \left(1 + k^2 \right)^s d^n k$$

where ^ denotes Fourier transform. Using the convolution theorem for Fourier transforms this is

$$A = \iiint \hat{f}(k + \ell_1) \hat{g}(\ell_1) \overline{\hat{f}}(k + \ell_2) \overline{\hat{g}}(\ell_2) \left(1 + k^2 \right)^s d^n k \, d^n \ell_1 d^n \ell_2$$

$$= \iiint f_1(k + \ell_1) f_4(\ell_1) f_2(k + \ell_2) f_3(\ell_2)$$

$$\phi(k, \ell_1, \ell_2) d^n k \, d^n \ell_1 d^n \ell_2 \qquad (7)$$

where

$$f_1(x_1) := \hat{f}(x_1) \left(1 + |x_1|^2 \right)^{t/2}$$

$$f_2(x_2) := \overline{\hat{f}}(x_2) \left(1 + |x_2|^2 \right)^{t/2}$$

$$f_3(x_3) := \overline{\hat{g}}(x_3) \left(1 + |x_3|^2 \right)^{u/2}$$

$$f_4(\ell_1) := \hat{g}(\ell_1) \left(1 + |\ell_1|^2 \right)^{u/2}$$

$$\phi(k, \ell_1, \ell_2) := \left(1 + |\ell_1 + k|^2 \right)^{-t/2} \left(1 + |\ell_1|^2 \right)^{-u/2}$$

$$\left(1 + |k|^2\right)^s \left(1 + |k + \ell_2|^2\right)^{-t/2} \left(1 + |\ell_2|^2\right)^{-u/2}.$$

Applying Cauchy's inequality to the (ℓ_1, ℓ_2)-integral in (7), regarded as an inner product between ϕ and $f_1 f_2 f_3 f_4$ in (ℓ_1, ℓ_2)-space, gives

$$A \leq \int d^n k \left[P_1(k) P_2(k) Q(k)\right]^{\frac{1}{2}} \tag{8}$$

with

$$P_1(k) = \int d^n \ell_1 \left[f_1(k + \ell_1)\right]^2 \left[f_4(\ell_1)\right]^2$$

$$P_2(k) = \int d^n \ell_2 \left[f_2(k + \ell_2)\right]^2 \left[f_3(\ell_2)\right]^2$$

$$Q(k) = \int \phi^2 d^n \ell_1 d^n \ell_2$$

Referring to (8), it is now sufficient to prove that Q is bounded and that $P_i^{1/2}(i = 1, 2)$ are L^2 functions, i.e. that the P_i are L^1 functions. Taking the latter task first, we have that

$$\int |P_1(k)| d^n x = \int d^n \ell_1 \left(\int d^n k \left[f_1(k + \ell_1)\right]^2\right) \left[f_4(\ell_1)\right]^2$$

$$= \|f\|_t \|g\|_u$$

as the required bound, and the same for P_2.

For Q we note that

$$Q(k)^{\frac{1}{2}} = B. \left(1 + |k|^2\right)^s$$

with

$$B = \int d^n \ell \left(1 + |\ell|^2\right)^{-u} \left(1 + |k + \ell|^2\right)^{-t}. \tag{9}$$

To evaluate B we divide $\mathbb{R}^n$ into the regions

$$X = \{\ell : |\ell| \leq |k|/2\}$$
$$Y = \{\ell : |\ell + k| \leq |k|/2\}$$
$$U = \{\ell : |\ell + k/2| \leq R\} \backslash X \backslash Y \qquad \text{where } R = (1 + |k|^2)^{\frac{1}{2}}$$
$$V = \mathbb{R}^n \backslash U \backslash X \backslash Y.$$

Denoting the integrand in the right hand side of (9) by h we have (since t is positive) that

$$\int_{X \cup U} h(k, \ell) d\ell \leq \int_{|\ell| < 2R} \left(1 + |\ell|^2\right)^{-u} \left(1 + |k|^2/4\right)^{-t} d\ell$$

$$\leq \text{const.} \times \left(1 + |k|^2\right)^{-t} \left(1 + R^{n-2u}\right)$$

$$\leq \text{ const.} \times \left[\left(1+|k|^2\right)^{-t} + \left(1+|k|^2\right)^{n/2-t-u}\right]$$

Similarly

$$\int_{Y \cup U} h\left(k, \ell\right) d\ell \leq \text{const.} \times \left[\left(1+|k|^2\right)^{-u} + \left(1+|k|^2\right)^{n/2-t-u}\right]$$

and

$$\int_V h\left(k, \ell\right) d\ell \leq \int_{|\ell'| \geq R} \left(1 + [|\ell'| - |k|/2]^2\right)^{-(u+t)} d^n \ell'$$

(putting $\ell' = \ell + k/2$)

$$\leq \text{ const.} \times \left(1+|k|^2\right)^{n/2-u-t}.$$

So

$$B \leq \text{const.} \times \left[\left(1+|k|^2\right)^{-t} + \left(1+|k|^2\right)^{-u}\right.$$
$$\left. + \left(1+|k|^2\right)^{n/2-u-t}\right].$$

Thus the conditions on s, t and u ensure that Q is bounded, as required. So we have a bound on the H^s norm of fg, and hence the product theorem is proved. $\qquad\square$

In the case of a space-like hypersurface in general relativity we have $n = 3$. Thus if f and g are both in H^s, then their product will be in H^s provided that $s > 1.5$. (In other words, the space H^s becomes furnished with a product and so becomes an algebra.) If f, g are derivatives of the metric, which are multiplied together in the expression for the Riemann tensor, then this requires the metric to be of differentiability $s > 2.5$.

For use in the next section we give another product theorem under more specialised assumptions.

Theorem 4.3.2 *Suppose that $\bar{\omega}$ is compact, that $g \in H^s$ with $0 < s < 1$ and that there exist positive constants K, α with $\alpha > s$ such that f satisfies the Hölder and boundedness conditions:*

$$|f(x) - f(y)| \leq K|x - y|^\alpha$$
$$|f(x)| \leq K \qquad\qquad (\forall x, y \in \bar{\omega})$$

Then $fg \in H^s$ with $\|fg\| \leq \text{const.} \times K\|g\|$.

Proof

Using the form (1) for the norm in H^s we can write

$$\|fg\|_s = \left\{ \iint |f(x)g(x) - f(y)g(y)|^2 |x - y|^{-n-2s} d^n x d^n y \right\}^{\frac{1}{2}}$$

$$\leq \left\{ \iint [|f(x)||g(x) - g(y)| \right.$$

$$\left. + |g(y)||f(x) - f(y)|]^2 |x - y|^{-n-2s} d^n x d^n y \right\}^{\frac{1}{2}}$$

$$\leq A + B$$

where

$$A = \left\{ \iint |f(x)|^2 |g(x) - g(y)|^2 |x - y|^{-n-2s} d^n x d^n y \right\}^{\frac{1}{2}}$$

$$B = \left\{ \iint |g(y)|^2 |f(x) - f(y)|^2 |x - y|^{-n-2s} d^n x d^n y \right\}^{\frac{1}{2}}$$

Now

$$A \leq \left(\sup_x |f(x)| \right)^2 \|g\|_s$$

while

$$B^2 \leq \int d^n y |g(y)|^2 \int d^n x K^2 |x - y|^{-n-2s+2\alpha}$$

(using the Hölder condition)

$$\leq \|g\|_s^2 K^2 \times \text{const.}$$

the constant depending on the diameter of the domain ω.

□

4.3.2 Functions of a function

We have noted that we also require the inverse of the metric to lie in the same function-space as the metric itself. Since the map from a symmetric matrix to its inverse is differentiable, indeed C^∞, where it is defined (by the implicit function theorem) this requirement is just a special case of the following result:

Theorem 4.3.3 *Let $0 < s$, $v = \lfloor s \rfloor$ so that $s = v + \sigma$ with $0 < \sigma < 1$, and let ω be a bounded domain in $\mathbb{R}^n$. Suppose that*

$$g \in H^s(\omega, \mathbb{R}^m) \cap C^k(\omega, \mathbb{R}^m)$$

for all natural numbers $k \leq \max(1, v/2)$, and
$$f \in C^{v+1}(\mathbb{R}^k, \mathbb{R}^\ell).$$
Then $f \circ g \in H^s(\omega, \mathbb{R}^\ell)$.

Proof As usual we begin by assuming that all our functions are C^∞ with support contained in ω, and obtain an estimate of the norms. From the definition (cf. (1))
$$\|f \circ g\|_s \leq \|D^v(f \circ g)\|_\sigma + \sum_{w \leq v} \|D^w(f \circ g)\|$$
(norms without a subscript denoting L^2 norms). The derivative $D^v(f \circ g)$ (and similarly $D^w(f \circ g)$) has the form
$$\|D^v(f \circ g)\|_\sigma = \left\| \sum_{r=1}^{v} ((D^r f) \circ g) \left(\sum_{\mathbf{k} \in P_r^v} \prod_i D^{k_i} g \right) \right\|_\sigma$$
where P_r^v denotes all partitions of the integers $\{1, \ldots, v\}$ into r non-empty subsets and, for $\mathbf{k} \in P_r^v$, k_i denotes the size of the ith subset. (The $\prod$ denotes a tensor product and the derivative $D^r f$, evaluated at a point obtained by composition with g, acts linearly on a sum of these tensor products.) Thus
$$\|D^v(f \circ g)\|_\sigma \leq \|((Df) \circ g)(D^v g)\|_\sigma$$
$$+ \sum_{r=2}^{v} \left\| ((D^r f) \circ g) \left(\sum_{\mathbf{k}} \prod_i D^{k_i} g \right) \right\|_1$$
$$\leq \|((Df) \circ g)(D^v g)\|_\sigma$$
$$+ \sum_{r=2}^{v} \left\| \left(D^{r+1} f \right) \circ g \right\|_{C^0} (\|g\|_{C^{[v/2]}})^{r-1} \|g\|_v$$
The terms after the summation signs are known to be bounded, and so it remains only to estimate the first term. But the H^σ-norm of the components of $((Df)_o g) \otimes (D^v g)$ are bounded by virtue of the second theorem in the last section, and so the result is proven. □

The restriction that g lies both in a Sobolev-space and in a C^k-space is rather artificial, although necessary for this pattern of proof. In most applications one can use a more natural condition by applying the Sobolev embedding theorem (Adams, 1975) which shows that $C^k(\omega) \subset H^s(\omega)$ (with a continuous inclusion – i.e.

the C^k-norm is bounded in terms of the H^s-norm) provided that $s > k + n/2$. So we can immediately conclude the following:

Corollary 4.3.4 . *The conclusions of the above theorem emain true if the conditions on g are replaced by the requirement that* $n > 2$ *and* $g \in H^s(\omega, \mathbb{R}^m)$ *provided that*

$$s > n \qquad (n \ even),$$

or

$$s > n - \frac{1}{2} \qquad (n \ odd).$$

In particular, if $n = 3$ (the case of a spacelike hypersurface in relativity) then we require $s > 2.5$. When applied to the formation of the inverse metric this therefore gives the same requirement as we obtained earlier when considering products.

4.4 Energy inequalities

Our aim in this section is not to give the details of the proofs of existence for solutions of Einstein's equations, but to explain why it is that these proofs lead one to the particular levels of differentiability that we have described. It is hoped that the account here will establish the basic concepts and techniques of the theory of partial differential equations so that the reader will be able to approach the papers proving the results without too much difficulty.

The proof of the existence of solutions to Einstein's equations proceeds in three steps. First, one makes a particular choice of the coordinates so that the equations take on a particularly tractable form. Usually one chooses harmonic coordinates, in which case the second derivatives of the metric components appearing in certain of the field equations simply constitute the wave operator acting on g_{ij}. Next one considers the linearisation of the resulting equation and shows that solutions exist to the linearised equations. Finally one uses theorems about non-linear equations in general to show that, if solutions exist to the linearised equations, then they also exist to the non-linear equations. Energy inequalities play a central role in the second of these steps, and it is here that the use of Sobolev spaces is decisive.

We first quote the form of the energy integral for linear hyperbolic systems given by (Hawking and Ellis, 1973) p. 237, to which the reader is referred for details of what follows. In order to write the result in a covariant form, these authors introduce a background metric $\hat{g}$ (of arbitrary signature) so as to define a connection and hence covariant derivatives, as well as a background positive definite metric e so as to define positive definite norms of tensors at a given point. If one chooses a particular coordinate system, working from now on in $\mathbb{R}^{n+1}$, then these can be suppressed by replacing covariant differentiation with respect to $\hat{g}$ by the partial derivative $(\ )_{,}$. The penalty for this is, of course, that one can only work locally, within a region sufficiently small for there to exist a single coordinate system valid throughout; but in fact the results are usually proved by local methods for other, technical reasons, so that this is no further restriction. We can take for e the Euclidean metric so that the norm of $K^I{}_J$, for example, is $e^{IM}e_{JN}K^J{}_I K^N{}_M = \sum (K^I{}_J)^2$.

Suppose that a tensor field $K^I{}_J$ (the capital indices abbreviating sets of tensor indices) satisfies a linear partial differential equation of the form

$$L(K) \equiv A^{ab} K^I{}_{J,ab} + B^{aPI}{}_{QJ} K^Q{}_{P,a} + C^{PI}{}_{QJ} K^Q{}_P = F^I{}_J \quad (9)$$

where A is a metric of Lorentz signature. Define

$$S^{ab} := \left\{ \left(A^{ac}A^{bd} -\frac{1}{2}A^{ab}A^{cd} \right) K^I{}_{J,c} K^P{}_{Q,d} - \frac{1}{2}A^{ab} K^I{}_J K^P{}_Q \right\} e^{JQ} e_{IP}.$$

Then the divergence of S is related to $L(K)$ and hence to F; more explicitly

$$S^{ab}{}_{,b} = F^P{}_Q A^{ac} K^I{}_{J,c} e^{JQ} e_{IP} + Q^a$$

where Q^a is a linear form with constant coefficients in the products

$$B \otimes A \otimes \partial K \otimes K, \quad C \otimes A \otimes K \otimes K, \quad A \otimes \partial A \otimes \partial K \otimes \partial K$$

$$\text{and} \quad \partial A \otimes K \otimes K.$$

Let U be an open set in $\mathbb{R}^n$ whose closure is a compact submanifold with boundary in a manifold foliated by spacelike hypersurfaces $H_t, (-\infty < t < \infty)$, t being a time coordinate, "spacelike" and "timelike" being defined relative to A, and consider the set

$$U_t = U \cap \bigcup_{0 < t' < t} H_{t'}$$

Then if we define $E(t)$ to be the energy integral

$$E(t) = \int_{H_t \cap U} S^{ab} t_{,a} t_{,b} dS$$

we see that this is a natural L^2-type measure of the size of K and its derivatives; namely we have

$$2E(t) = \int_{H_t \cap U} \left(h^{ab} K^I{}_{J,a} K^P{}_{Q,b} \right.$$
$$\left. + K^I{}_J K^P{}_Q + \dot{K}^I{}_J \dot{K}^P{}_Q \right) e^{JQ} e_{IP} dS \quad (10)$$

where h^{ab} is the Riemannian metric induced by A on H_t and $\dot{}$ denotes covariant differentiation with respect to t^a.

It is simple to convert the terms in this expression into actual L^2 norms of the form used in our previous discussion. Consider first the expression

$$I_h = \int_\omega h^{ab} u_a u_b d^{n-1} x$$

for the size of a vector field over H, such as occurs in (10), and compare it with the L^2 norm

$$\int_\omega e^{ab} u_a u_b dS = \|u\|_0$$

where e is the Euclidean metric. Since h is positive definite and symmetric, there is a unique symmetric positive definite metric k $(= h^{\frac{1}{2}})$ such that $k^{ab} k^{cd} e_{bc} = h^{ad}$. Hence

$$\|u\|_0 = \int e^{ab} k^{-1}{}_{af} k^{fc} u_c k^{-1}{}_{bm} k^{md} u_d d^{n-1} x$$
$$= \int h^{-1}{}_{fm} k^{fc} k^{md} u_c u_d d^{n-1} x$$
$$\leq \| h^{-1} \|_s \|ku\|_0{}^2$$
$$= \|h\|_s I_h$$

where we have used the product theorem of section 4.3.1, valid provided that $s > n/2$. Moreover, the function-of-a-function theorem (4.3.3) allows us to estimate the norm of h^{-1} under the same conditions.

We also have that

$$I_h \leq \int \|h(x)\| \|u\|^2 d^{n-1}x$$

$$\leq \|h\|_s \|u\|_0^2$$

under the same conditions.

Thus if u is the derivative of a function ϕ then $\|u\|_0 = \|\phi\|_1$; moreover, the argument is equally applicable to vector valued functions.

So applying this idea to the function K in the definition of E we obtain

$$C_1\left(\|K, K\|_{1,0}\right)\|h\|_s \leq E(t) \leq C_2\left(\|K, K\|_{1,0}\right)\|h\|_s \tag{11}$$

for constants C_1, C_2 , with $\| \|_{1,0}$ denoting the norm

$$\|K, L\|_{1,0} = \|K\|_1 + \|L\|_0$$

on $H^1 \times H^0$.

Let ∂U_t be the boundary of U_t (piecewise differentiable) and n its normal. Clearly ∂U_t comprises parts in H_0 and in H_t, together with the remainder, in the boundary of U; we suppose this remainder to have a timelike or null future-pointing normal n.

Applying the divergence theorem to this set gives

$$\int_{U_t} \left(S^{ab}t_{,a}\right)_{,b} dv = \int_{U_t} S^{ab}{}_{,b}t_{,a} dv + \int_{U_t} S^{ab}t_{,ab} dv$$

$$= \int_{\partial U_t \cap \partial U} S^{ab}t_{,a}n_b dS - E(0) + E(t) \tag{12}$$

Now it is easily seen that if a and b are two future-pointing timelike vectors (so that $A^{ij}a_i b_j < 0$) then

$$S^{ij}a_i b_j \geq 0. \tag{13}$$

Applying this to the first term in (12) gives

$$\int_{U_t} S^{ab}{}_{,b}t_{,a} dv + \int_{U_t} S^{ab}t_{,ab} dv \geq -E(0) + E(t) \tag{14}$$

The first term here is, as we have seen, given via $L(K)$. And the condition (13) (called the dominant energy condition) implies that, in any orthonormal basis (with respect to A), $S^{00} \geq |S^{ab}|$ for each a, b. In other words, the whole size of S is estimated by S^{00} in a given basis.

Now, if A is continuous (which will be the case, by Sobolev's embedding theorem, if it is of class H^s with $s > n/2$) then we can choose the slicing by hypersurfaces to be differentiable and then, by smoothing (Seifert, 1977), C^∞. Thus the components of $t_{,ab}$ will be bounded above on H_t, as will the A-norm of $t_{,a}$. Consequently we have that

$$S^{ab}t_{,ab} \leq PS^{ab}t_{,a}t_{,b}$$

for a positive constant P, on H_t. So (14) then takes the form

$$E(t) \leq E(0) + \int \left(\int_{H_{t'} \cap U} S^{ab}{}_{,b}t_{,a}dS + P \int_{H_{t'} \cap U} S^{ab}t_{,a}t_{,b}dS \right) dt'$$

$$= E(0) + \int \left(\int \left[F^P{}_Q A^{ac} K^I{}_{J,c} e^{JQ} e_I{}_P t_{,a} + Q^a t_{,a} \right] dS \right.$$
$$\left. + PE(t') \right) dt'$$

giving an estimate of the size of K at t in terms of the size at $t = 0$ and in terms of what goes on in between 0 and t.

In order to estimate the size of the terms in the integral, suppose that $A \in H^{1+s}$, $B \in H^s$ and $C \in H^{s-1}$. (Actually in the case of general relativity we know that $C \in H^s$). So we can define

$$N = \sup_{0 < t' < t} \max \left[\|A\|_{1+s}, \|B\|_s, \|C\|_{s-1} \right].$$

If $s > \max(n/2, 3n/4 - 1)$ then the product theorem can be applied to the product FAK in the preceding integral and to the products occurring in the specification of Q. The condition $s > n/2$ ensures that ∂A and A are bounded, and hence bounds the terms in Q, while $s > (3n/4 - 1)$ (a condition that is only stronger than $s > n/2$ for $n \geq 4$) takes care of the F term.

For then applying (11) gives

$$E^*(t)^2 \leq E^*(0)^2 + \int N \left[\|F\|_s E^*(t') + PE^*(t')^2 \right] dt'$$

where

$$E^*(t) = \|K, \dot{K}\|_{1,0}$$

(the norms, as always here, being evaluated over the space-like hypersurface $H_t \cap U$ of dimension $n - 1$).

The solution to this integral inequality is easily seen to be

$$E^*(t) \leq (E^*(0) + N/P) e^{Pt},$$

which is the basic energy estimate for this hyperbolic equation.

To extend this estimate to higher derivatives, the simplest approach is to differentiate the differential equation, and put

$$K^P{}_{Qc} = K^P{}_{Q,c}$$

to give

$$A^{ab}K^I{}_{Jc,ab} + \left(B^{aPI}{}_{QJ}\delta^b{}_c + A^{ab}{}_{,c}\delta^I{}_Q\delta^P{}_J\right)K^Q{}_{Pa,b}$$

$$+ \left(C^{PI}{}_{QJ}\delta^a{}_c + B^{aPI}{}_{QJ,c}\right)K^Q{}_{Pa} = F^I{}_{J,c} - C^{Pi}{}_{QJ,c}K^Q{}_P \quad (15)$$

an equation of the same form as before for $K^Q{}_{Pc}$ but with K and the derivative of C appearing as a source. We can repeat the former argument, providing that $C \in H^{s+n/2-1}$ (in order to ensure that $\partial C \otimes K \in H^{s-1}$, taking account of the fact that we already know $K \in H^1$) and that $F \in H^s$. If these conditions are met, then we can show that $K \in H^2$.

The problem is, that in the application we have in mind to general relativity this condition on C is not met: C involves the connection and so if $g \in H^{s+1}$ then all we know is that $C \in H^s$. So in order to increase the level of differentiability in the energy estimate we are not able to differentiate the equation even once: it is necessary to differentiate it a "fractional number of times"; and of course, something like this will inevitably be needed if one is to obtain an energy estimate in H^{s+1}.

The solution is to use the form of the norm on H^s involving convolution with a Bessel potential (4.2.2). We note first that the reason for its being possible to apply the same analysis to (15) as to (9) is that the highest order terms are the same, and the other terms are of the same differentiability in the two equations. Expressed in other words, the difference between $\partial_c LK$ and $L\partial_c K$ does not involve second derivatives; i.e. the commutator $[\partial_c, L]$ is a first order differential operator, with terms of the appropriate differentiability. The same is true of the the commutator $[G*, L]$, where $G*$ is the operation of convolution with the Bessel potential G, and it is this property of the commutator (expressed in abstract terms) that is the key assumption in the theorems of Kato (1975) on which the current existence theorems are based. To see this, we need only note that, from its definition in terms of Fourier transforms, $G*$ commutes with derivatives; while if we consider, for example, the commutator of $G*$ with the second order differential operator $A^{ab}\partial_a\partial_b$, then

$$G * A^{ab}\partial_a\partial_b K - A^{ab}\partial_a\partial_b G * K$$
$$= \int \left[A^{ab}(y) - A^{ab}(x)\right] G(x,y)\partial_a\partial_b K.$$

4.5 Linking equations and geometry

The condition that the metric satisfy a Sobolev condition on certain spacelike surfaces is very unsuited to geometric constructions. Roughly speaking, geometry does not respect the division into space and time that is implied in the reduction of Einstein's equations by the use of harmonic coordinates, as in the proof of the existence of solutions. Instead, it will be necessary for us to work with Sobolev conditions defined on the whole of space-time, i.e. with conditions such as the requirement that each component g_{ij} should belong to $H^s(\mathbb{R}^4)$. But we should obviously aim to choose these conditions to be as close as possible to the conditions involved in the existence theorems that are at present known, in the hope that in the future an exact match might be achieved between geometrical and analytical conditions.

Although the result of Hughes et al. (1977) refers to the Sobolev class on the surfaces $t =$ const. in a particular (Harmonic) coordinate system, it is an integral part of the techniques used – specifically, the energy inequalities – that these conditions on the metric will still hold if one does a "reasonable" transformation of the coordinates. Indeed, one would expect physically that the conditions would be at least stable under small variations of the hypersurfaces $t =$ const. In other words, the conditions still hold not only for one special slicing but for all smooth coordinate systems x' having $dx^{0\prime}$ sufficiently close to dx^0. Under larger variations the conditions could break down through the surfaces becoming tangent to characteristics of the evolution equations.

The approach to be taken uses the fact that, if a function f (specifically, a component of the metric or of the Riemann tensor) with compact support defined on $\mathbb{R}^{n+1}$ (regarded as a Minkowski space with x^0 timelike) satisfies a Sobolev condition on all "sufficiently spacelike" hypersurfaces, then it satisfies the condition (at the same level of differentiability) on the whole space. (The converse is not true.) This can then be used to allow one to deduce the differentiability of the metric or the Riemann tensor, as a local Sobolev class on the whole space-time, given knowledge of its behaviour on spacelike hypersurfaces.

More precisely, we have:

Proposition 4.5.1

Suppose that there exists (locally) a positive number ϵ such that whenever $\mathbf{n}$ is a unit vector restricted to

$$n^0 > 1 - \epsilon \tag{16}$$

and t a real number, if we define the surface

$$S_{t,\mathbf{n}} = \{x \mid x^0 = t + x.\mathbf{n}\}$$

then f satisfies

$$f|S_{t,\mathbf{n}} \in H^s(S_{t,\mathbf{n}}). \tag{17}$$

Under these circumstances, $f \in H^s(\mathbb{R}^4)$.

(The notation $f \in H^s(\Sigma)$ where Σ is a surface
$$\{(t,x,y,z)|t = \sigma(x,y,z)\}$$
means that $\bar{f} \in H^s(\mathbb{R}^3)$ where $\bar{f}(x,y,z) := f(\sigma(x,y,z),x,y,z)$.)

Proof

It is sufficient to consider the case where $0 < s < 1$, since other non-integral s can be handled by applying the same arguments to the derivatives of f. Then we shall first show that f has a finite value of the modified norm

$$\left\{ \sum_{|\mathbf{i}|=m} \iint_Q |D^{\mathbf{i}} f(x) - D^{\mathbf{i}} f(y)| / |x - y|^{n+2s} dx dy \right\}^{\frac{1}{2}} = \|f\|_Q$$

where Q is the domain of all pairs (x, y) such that for some $t, \mathbf{n}$ as above x and y both lie on $S_{t,\mathbf{n}}$. This is equivalent to

$$(x - y)^0 / |x - y| < \epsilon.$$

For, condition (17) gives the finiteness of

$$\iint d\mathbf{n} dt \|f\|_s = \iint dx dy \iint d\mathbf{n} dt |f(x, t - x.\mathbf{n}) - f(y, t - y.\mathbf{n})|$$
$$\left\{ (x - y)^2 + [(x - y).\mathbf{n}]^2 \right\}^{-(n+2\sigma)/2} \tag{18}$$

(from the definition (1) of the Sobolev norm), where the domain of integration for n is the set of all unit vectors satisfying (16), with measure

$$\mathbf{dn} = dn^1 dn^2 \ldots dn^n.$$

We define $\bar{x}, \bar{y}, \mathbf{n}'$ by imposing the equations

$$(x, t - x.\mathbf{n}) = \bar{x}$$
$$(y, t - y.\mathbf{n}) = \bar{y}$$
$$\mathbf{n} = \mathbf{n}^*/|\mathbf{n}^*|$$
$$\mathbf{n}^* = \mathbf{n}' - [\mathbf{n}'.(\bar{x} - \bar{y})/e.(\bar{x} - \bar{y})]e$$

with $\mathbf{e}$ a fixed unit vector having $e^0 = 0$ and $\mathbf{n}'$ restricted to

$$\mathbf{n}'^0 = 1, \qquad \mathbf{n}'.e = 0.$$

These define a smooth 1-to-1 correspondence between the old $(x, y, \mathbf{n}, t)$ and new $(\bar{x}, \bar{y}, \mathbf{n}')$ variables within a neighbourhood of the points where

$$(\exists k) \left((\bar{x} - \bar{y})^i = ke^i, i = 1, \ldots, n \right). \tag{19}$$

To use these new variables of integration in (18) we need the Jacobean of the transformation thus defined, i.e.

$$\det \begin{bmatrix} \frac{\partial x}{\partial \bar{x}} & \frac{\partial x}{\partial \bar{y}} & \frac{\partial x}{\partial \mathbf{n}'} \\ \frac{\partial y}{\partial \bar{x}} & \frac{\partial y}{\partial \bar{y}} & \frac{\partial y}{\partial \mathbf{n}'} \\ \frac{\partial t}{\partial \bar{x}} & \frac{\partial t}{\partial \bar{y}} & \frac{\partial t}{\partial \mathbf{n}'} \end{bmatrix} \tag{20}$$

where the entries are matrices. Clearly

$$\frac{\partial x}{\partial \bar{x}} = [I_n \quad \mathbf{0}], \qquad \frac{\partial x}{\partial \bar{y}} = O$$

and similarly with x and y interchanged, while

$$\frac{\partial n^i}{\partial \bar{x}^\nu} = -\frac{\partial n^i}{\partial \bar{y}^\nu} = A^i{}_\nu, \quad \text{say}.$$

If we write the relation between t and the new variables as

$$t = \bar{x}^0 + n_i(\bar{x}, \bar{y}, \mathbf{n}')\bar{x}^i \qquad (i = 1, \ldots, n, \text{ summed})$$

then

$$\frac{\partial t}{\partial \bar{x}^0} = 1 + \frac{\partial n^i}{\partial \bar{x}^0}\bar{x}_i$$

and

$$\frac{\partial t}{\partial \bar{y}^0} = -\frac{\partial n^i}{\partial \bar{x}^0}\bar{x}_i;$$

Inserting these into (20) we find that the determinant can be reduced to

$$\det \begin{bmatrix} I_n & \mathbf{0} & O_n & \mathbf{0} & O_n \\ O_n & \mathbf{0} & I_n & \mathbf{0} & O_n \\ O_n & \mathbf{0} & O_n & -A^i{}_0 & P \\ \mathbf{0}^t & 1 & \mathbf{0}^t & 0 & \mathbf{0}^t \end{bmatrix} = \det \left[A^i{}_0 P \right]$$

($\mathbf{0}$ denotes the zero column vector, t denotes transpose, O_n the $n \times n$ zero matrix, I_n the $n \times n$ unit matrix and $P = \partial n / \partial \mathbf{n}'$).

For $A^i{}_0$ we calculate

$$A^i{}_0 = -\frac{\alpha^2}{(\alpha^2 + \beta^2)^{3/2}} \left[e^i + \frac{\beta}{\alpha} \mathbf{n}'^i \right]$$

where $\alpha = e.(\bar{x} - \bar{y})$ and $\beta = \mathbf{n}'.(\bar{x} - \bar{y})$.

The calculation of P is simplified if we are at a point where (19) holds and if we also choose coordinates so that $e = (1, 0, ..., 0)^t$. Then

$$\frac{\partial n^i}{\partial n'^A} = \frac{1}{|n^*|} \delta^i_A - \frac{\mathbf{n}'_A}{|n^*|^3} \left(n'^i - \frac{\beta}{\alpha} \delta^i_1 \right) \qquad (A = 2, 3, \ldots, n).$$

We see now that all the functions involved in the determinant are bounded (globally) except for the factor $\alpha^2 / (\alpha^2 + \beta^2)^{3/2}$. If we take out this factor then the determinant that remains, namely

$$\det \left[\left(e^i + \{\beta/\alpha\} n'^i \right) P \right]$$

is easily shown to be bounded below. Now from the definitions of α and β

$$\frac{\alpha^2}{(\alpha^2 + \beta^2)^{3/2}} |\bar{x} - \bar{y}|$$

is bounded above and below. So if we pass to the new variables and perform the $\mathbf{n}'$-integration, we aquire an extra factor of $1/|\bar{x} - \bar{y}|$ from the Jacobean; and so we obtain the expression for $\|f\|_Q$.

To pass from $\|f\|_Q$ to $\|f\|_s$ we write (1) (still keeping to the case where $0 < s < 1$) as

$$\|f\|_s^2 \leq I_1 + I_2,$$

where

$$I_1 = \iint |f(x) - f(z(x,y))| \, |x - y|^{-2n-2s} dx dy$$

$$I_2 = \iint |f(z(x,y)) - f(y)| \, |x - y|^{-2n-2s} dx dy$$

$$z(x,y) = \left(\frac{x+y}{2} + K|x-y|e\right)$$

with e a fixed unit vector.

If we choose K large enough, the integrand in I_1 is zero unless (x,z) lies in Q, and the same with (y,z) in the case of I_2. The Jacobean of the transformation from (x,y) to (x,z) in I_1 and from (x,y) to (z,y) in I_2 is bounded, and so both these integrals can be bounded by $\|f\|_Q$ as required. $\qquad\square$

Definitions. This result, together with the existence theorems for Einstein's equations discussed in the last section, justifies us in taking as our smallest practicable differentiability criterion the condition that in some atlas (namely, one smoothly related to the harmonic coordinate system used in the proof of the existence theorems) the components of the metric are locally of class $H^{2.5+\epsilon}$ for some positive ϵ. Thus we shall speak of an H^s space-time (with a given atlas) if the components of the metric are locally of class H^s in all charts of the given atlas.

If we incorporate the idea of formulating differentiability in terms of the Riemann tensor, then we reach the following definition. A space-time will be called of class $\mathbf{H}^s$ if it satisfies (i) of section 4.2.1 and if, in addition, the components of the Riemann tensor are locally in H^s.

4.5.1 Discussion

It must be borne in mind that conditions $H^{2.5+\epsilon}$ and $\mathbf{H}^{0.5+\epsilon}$ are weaker than the condition used in the existence theorem: the existence theorem levels imply the Sobolev class on $\mathbb{R}^4$ but not vice versa. Mathematically, the difference lies in the fact that the existence theorems deal with classes on hypersurfaces where, because of the lower dimension, the condition that the connection coefficients are H^s for $s > 1.5$ implies that they are bounded. (We have already seen that this is crucial for allowing one to multiply these coefficients when considering the Einstein equations.) As a result, we have to impose boundedness of these coefficients as a separate condition (4.2.1 (i)). It is this that will prevent us from carrying all our results through into the Sobolev case. For it will turn out that the operations that preserve the H^s condition on the Riemann tensor when constructing an extension of the space-time do not preserve the boundedness of the connection.

For this reason existence of solutions to Einstein's equations cannot be assumed at this level. In order to obtain existence from the presently existing theorems we have to raise the levels of differentiability by 0.5, which ensures than the differentiability on spacelike hypersurfaces is $H^{2.5+\epsilon}$. It might, however, be hoped that in the future an existence theorem might be available for Sobolev classes on $\mathbb{R}^4$; probably in the form of a solution to an "initial slab" problem, rather than an initial surface problem, in which one was given a solution to the Einstein equations in a globally hyperbolic region with spacelike future boundary and the existence of a continuation was shown.

4.6 Bounds on the metric in particular charts

It is not easy to relate approaches based on the differentiability of the Riemann tensor to those which use the metric components. It is clear from the definition of the Riemann tensor that, if the metric is C^{X+1} then the Riemann tensor will be C^{X-1}. But it is not known whether or not the converse is true. In the present section we shall derive for the Hölder differentiability classes the weaker result that if the Riemann tensor is C^{X-1} then there are coordinates in which the metric is C^X. Results of this kind fail in the Sobolev case because of the problem of the boundedness of the connection discussed in the previous section.

In deciding on whether or not a space-time constructed in a particular way is C^X, a useful technique is to apply a process of "smoothing" to the metric. This will be explained below in section 4.8; but the idea is to approximate the metric by a sequence $\{g_n\}$ of C^∞ metrics tending to the given metric as a limit. If the components of the metrics in the sequence, or of their derivatives, are not bounded as $n \to \infty$ then the limiting tensor (or derivative) will not exist everywhere in the space-time, and the converse can be made to hold under suitable additional restrictions. Thus the basic problem becomes one of determining whether or not various derivatives of a smooth metric can be given bounds in terms of the Riemann tensor, whose bounds are supposed known.

The basic formulae used for estimating the size of the derivatives of the metric in a given coordinate system are easily written down. We have

$$g^{ij} = \overset{-1}{g}\left(dx^i, dx^j\right)$$

and so

$$g^{ij}{}_{,k} = 2\overset{-1}{g}\left(\nabla_{\partial_k} dx^{(i}, dx^{j)}\right).$$

This in turn determines the derivative of the covariant metric as

$$g_{ij,k} = -g_{i\ell}g^{\ell m}{}_{,k}g_{mj} \tag{22}$$

In this way the problem of estimating the size of $g_{ij,k}$ is reduced to that of estimating $\nabla_U dx^i$ in the case $U = \partial/\partial x^k$.

4.6.1 Normal coordinates

We first illustrate the procedure for the case of normal coordinates about a point x_0 with respect to a pseudo-orthonormal frame $\mathbf{E}$ at x_0. Although this is geometricaly the most natural coordinate system, it turns out to be inadequate for our purposes here. The method of defining such coordinates is to draw from x_0 a geodesic $\lambda_x(s)$ to every point x in a normal coordinate neighbourhood of x_0 parametrised by

$$\lambda_x(1) = x, \qquad \lambda_x(0) = x_0.$$

The normal coordinates of the point x are then defined by

$$z^i(x) = -\dot\lambda_x(0).\overset{i}{E} \tag{23}$$

where "." denotes the Lorentz inner product and the index on E is raised by the flat-space metric.

Next, to find dz^i draw an arbitrary curve $\mu : t \mapsto \mu(t)$ with $\mu(0) = x$, and consider the equation

$$dz^i\left(\dot\mu(0)\right) = \frac{d}{dt}z^i\left(\mu(t)\right)\Big|_{t=0} \tag{24}$$

Define $\bar\lambda(t,s) := \lambda_{\mu(t)}(s)$ and set

$$\bar\lambda_*\left(\frac{\partial}{\partial t}\right)\Big|_{(t,s)} =: \bar{Y}(t,s)$$

so that $\bar{Y}(0,1)) = \dot{\mu}(0)$ and $\bar{Y}$ is a Jacobi field on the family $\lambda_{\mu(t)}$ of geodesics. That is, setting

$$X := \lambda_* \left(\frac{\partial}{\partial s} \right) = \dot{\lambda}_{\mu(t)}$$

we have

$$\nabla_{\bar{Y}} X = \nabla_X \bar{Y} \qquad (25)$$

$$\nabla_X \nabla_X \bar{Y} = R(X, \bar{Y})X. \qquad (26)$$

Writing $Y := \bar{Y}(0,1) = \dot{\mu}(0)$, (23) and (24) then give

$$dz^i(Y) = - \left(\nabla_{\bar{Y}(0,0)} X \right) \overset{i}{.} E = - (\nabla_X \bar{Y}) \Big|_{s=0} \overset{i}{.} E. \qquad (27)$$

Now parallely propagate the basis vectors $\mathbf{E}$ along the geodesics λ_x , thereby extending $\mathbf{E}$ to a field satisfying $\nabla_X \mathbf{E} = 0$. Taking components with respect to this extended $\mathbf{E}$ gives (27) as

$$dz^i(Y) = - \frac{dY^i}{ds} \Big|_{s=0} \qquad (28)$$

while (26) becomes

$$\frac{d^2 Y^i}{ds^2} = R^i{}_{jkl} X^l X^j Y^k. \qquad (29)$$

(This is just the Jacobi equation already derived in chapter 2).

Integrating from 0 to s gives

$$\frac{dY^i}{ds} = \frac{dY^i}{ds} \Big|_0 + \int R^i{}_{jkl} X^l X^j Y^k ds;$$

and integrating again from 0 to 1 gives

$$-Y^i = -Y^i(0) = \frac{dY^i}{ds} \Big|_0 + \iint R^i{}_{jkl} X^l X^j Y^k ds' ds.$$

Eliminating dY^i/ds and using (28) thus gives

$$dz^i(Y) = Y^i + \int_0^1 ds \int_0^s Q ds' - \int_0^1 Q ds$$

where Q is the integrand of the preceding equations.

Thus we see that the size of $dz^i(Y)$, and hence the size of dz^i (since Y was the tangent to an arbitrary curve) depends on the size of the Riemann tensor. A repetition of this argument will then show that the size of $\nabla_U dz^i$, and hence of $g_{ij,k}$, depends on the derivative of the Riemann tensor. So, even if the Riemann tensor itself is bounded, there is no guarantee that the derivative of the metric components in normal coordinates will exist, unless we also have information about the derivatives of the Riemann tensor. For this reason normal coordinates are not very useful if our aim is to achieve a differentiable metric.

4.7 Distance-function coordinates

In section 6.1 of (Clarke, 1982) it is shown that we can do better by taking as coordinates the quantities $w^i(x) = d(p^i, x)$, where the p^i are four suitably positioned points and we define $d(x, y)$, in the case where x is to the past of y and both are contained within a single normal coordinate neighbourhood, as the proper time along a geodesic connecting x to y. We shall call these chronological coordinates.

Then it turns out that the metric is given by

$$g^{ab} = g\left(\overset{a}{Z}, \overset{b}{Z}\right)$$

where the $\overset{a}{Z}$ are vector fields defined by the requirement that $\overset{a}{Z}(p)$ is the tangent vector to the geodesic from p to the point p^a, normalised by $g(\overset{a}{Z}, \overset{a}{Z}) = -1$. In this case a procedure like that above gives an expression for the derivatives $g_{ij,k}$ in which only the Riemann tensor appears and not its derivative. This seems to be the best that can be achieved, in the sense that there does not appear to be any direct extension of these methods which would give $g_{ij,kl}$ in terms of the Riemann tensor, with no derivatives of it.

The chronological coordinates w^i have, however a disadvantage that is rather analogous to what occurs in polar coordinates. In the latter r is a Euclidean distance and there is a coordinate singularity with g becoming large as $r \to 0$. Correspondingly, the smaller the coordinates w the larger is the norm of the $g_{ij,k}$. (In this case one cannot speak of a coordinate singularity as $w \to 0$ since the coordinates are not defined up to any points $w^i = 0$.)

To circumvent this problem we shall define coordinates z^i (different from the normal coordinates in the previous section) to be $d(x, S^i)$, where S^i is a surface. If the S^i are fairly smooth this will remove the "polar coordinate" behaviour. To achieve the smoothest S^i we shall take S^i to be the surface equidistant between two points, p^i and q^i.

For the time being we shall just study a single coordinate and so we shall omit the "i" from z^i, p^i and so on. Thus we define

$$S = \{x \in I^+(p) \cap I^-(q) : d(p, x) = d(x, q)\}.$$

4.7.1 The normal to S

In order to find an expression for the normal, suppose we have a curve κ in S, $\kappa : [0, 1] \to S$. Set γ_{pt} for the geodesic from p to $\kappa(t)$ with $\gamma_{pt}(0) = p$, $\gamma_{pt}(1) = \kappa(t)$ and similarly for q. Write the tangent vectors to these geodesics as $V_{pt} := \dot{\gamma}_{pt}$ and define the corresponding unit vectors by

$$Z_{pt} = \frac{V_{pt}}{|V_{pt}.V_{pt}|^{1/2}}.$$

We make corresponding definitions with q replacing p.

It is now simple to show that the normal to S at $\kappa(t)$ is $Z_{pt} - Z_{qt}$. Indeed, the fact that κ lies in S means that

$$\frac{d}{dt}d(p, \kappa(t)) = \frac{d}{dt}d(\kappa(t), q). \tag{30}$$

But

$$\frac{d}{dt}d(p, \kappa(t)) = \frac{d}{dt}|V_{pt}.V_{pt}|^{1/2} = -|V_{pt}.V_{pt}|^{-1/2}V_{pt}.\nabla_W V_{pt}$$

(where we write $W(t)$ for the value on S of the connecting vector field $\bar{W}_p = \gamma_{pt*}(\partial/\partial t)$ namely $W(t) = \bar{W}_p(1, t) = \dot{\kappa}(t)$), i.e.

$$\frac{d}{dt}d(p, \kappa(t)) = -\left(\frac{1}{d(p, \kappa(t))}\right) V_{pt}.\nabla_{V_{pt}}\bar{W}$$

$$= -\left(\frac{1}{d(p, \kappa(t))}\right) \nabla_{V_{pt}}(V_{pt}.\bar{W})$$

$$= -\frac{V_{pt}.\bar{W}}{d(p, \kappa(t))}$$

(since $(d^2/ds^2)V.\bar{W} = 0$, $V.\bar{W}|_{s=0} = 0$)

$$= -W.Z_{pt}.$$

So from (30)

$$W.Z_{pt} = W.Z_{qt}$$

or

$$W.(Z_{pt} - Z_{qt}) = 0.$$

But we can draw κ so as to obtain an arbitrary vector tangent to S as W, and so this means that the above relation holds for all tangent vectors to S, in other words $Z_{pt} - Z_{qt}$ is proportional to the normal. Finally, since V_{pt} and V_{qt} are of the same length, this means that $V_{pt} - V_{qt}$ is also proportional to the normal.

4.7.2 The curvature of S

The smoothness of the coordinates being constructed will depend on the smoothness of S, which can be expressd in terms of the derivative of the unit normal N (equivalent to the extrinsic curvature of S). To calculate this let $N' = V_p - V_q$ (now omitting the subscript t where no confusion would result) with κ a curve in S with tangent vector W, as in the previous section. In order to calculate the derivative of N' we need the derivatives of V_p and V_q, for which we use the Corollary 2.1.3 with V replacing Y and W replacing X. In the notation of that section we can write

$$\nabla_W V_p = W + Q_p \tag{31}$$

where, from 2.1.3,

$$|Q_p| = |\dot{W} - W| < 1.32 r_0 |W||Z_p|^2 \tag{32}$$

and where here, and throughout this section, we refer all norms to a field of orthonormal frames obtained by parallel propagation from p (or from q, remembering that these two possibilities are in general not compatible).

It follows that

$$|\nabla_W N'| < 2.64|W|r_0|V_p|^2 \tag{33}$$

We now assume that we are working in a convex normal neighbourhood so that p and q can be joined by a geodesic γ with tangent vector X_0, normalised so that $g(X_0, X_0) = -x_0$, where $x_0 = d(p, q)$. Furthermore, we assume that the orthonormal reference frames are defined by choosing a frame at p in which X_0 is the timelike member of the frame, so that $|X_0| = x_0$. If γ cuts S in the point n and we draw κ so that $n = \kappa(0)$, then we have that

$$V_{p0} = X_0/2, \quad V_{q0} = X_0/2$$

and so

$$N'(n) = X_0.$$

In order to calculate the rate of change of vectors along κ we use a different frame $\mathbf{F}$ that is parallely propagated along κ. Then $\mathbf{E}$ and $\mathbf{F}$ will be related as described in section 2.2. If we work within a small enough neighbourhood then we can use the results of that section, principally 2.2.5, to relate components in the two frames. That subsection gives an explicit expression for the length of the curves involved in terms of the Riemann tensor components.

Let T be any vector field defined along κ. Then if T' denotes components with respect to $\mathbf{F}$, we have

$$T^i(\kappa(t)) - T^i(\kappa(0)) = L_j{}^i \int L^j{}_k (\nabla_W T)^k \, dt + T^j(\kappa(0)) \Delta L_j{}^i \quad (34)$$

where

$$\Delta L^i{}_j = L^i{}_j - \delta^i{}_j$$

It will be convenient to reparametrise κ so that $|W(t)| = w$, a constant. In accordance with the notation of section 2.2.5 we shall put

$$k := \max_t |V_{pt}|$$

and impose the requirements

$$r_0 k^2 < 0.5 \quad (35)$$

$$r_0 k w < 0.1. \quad (36)$$

The first of these implies

$$r_0 k^2 < \pi^2/4$$

as is required for the results of 2.2 to hold. Then from that section we have

$$\|L\| < 1.4$$
$$\|\Delta L\| < 4.61 r_0 k w < .461$$

so that (34) becomes

$$|T(t) - T(0)| < 2 \int |\nabla_W T| \, dt + 4.61 r_0 k t w |T(0)| \quad (37)$$

Applying this to V_p we obtain from (31) and (32)

$$|V_p(t)| < |V_p(0)| + 2tw + 2.64r_0tk^2w + 4.61r_0kt|V_p(0)|w$$
$$< |V_p(0)|(1 + 0.461t) + 2.82wt \qquad (38)$$

The pattern of argument used in 2.2.5 can now easily be adapted to show that this will be achieved if instead of (35) and (36) we impose

$$r_0x_0/2 \quad < \quad 0.2$$
$$r_0w \quad < 0.14$$

which can be shown to imply (35), (36) and (38). For simplicity we shall work with $w < x_0$, $x_0 < 0.14r_0{}^{-}1/2$, so that only the size of x_0 need be controlled. Then from (38) we have

$$k < 3.55x_0 \qquad (39)$$

We now apply (37) to N' using (33) and (39) to obtain

$$|\Delta N'| = |N'(t) - N'(0)| < 67x_0{}^2tr_0 + 3.55x_0{}^2r_0t|N'(0)|$$

from which

$$|N'|/|N'(0)| = |N'|/x_0 < 83x_0{}^2tr_0.$$

Thus for $x_0\sqrt{r_0}$ sufficiently small we can have

$$|\Delta N'| < |N'(0)|/2,$$

and so the *metric* length of N', which we denote by $d(N')$, will satisfy $d(N') > d(N'(0))/2$, and $|N'|/d(N') < 2$. Thus setting $N = N'/d(N')$ we finally reach

$$|\nabla_W N| < |\nabla_W N'| \left(\frac{1}{d(N')} + \frac{|N'|^2}{d(N')^3} \right)$$
$$< 134x_0wr_0$$

from (33), and using the fact that we have chosen tetrads so that $d(X_0) = x_0$.

The second fundamental form ω of S (i.e. its extrinsic curvature) is obtained by projecting the tensor ∇N into the tangent space to S, and so

$$|\omega| < 134x_0r_0|N|$$
$$< 201x_0r_0.$$

4.7.3   Maximal geodesics

We next find the size of the neighbourhood of S in which there are maximal geodesics to S. It is well known that such a neighbourhood exists provided that S is achronal, as is the case here.

So, suppose that there exist two different geodesics γ_1 and γ_2 from a point x to S, both maximal and hence both cutting S perpendicularly in points y_1 and y_2 respectively. Let κ be the geodesic in S (regarded as a Riemannian submanifold) joining y_1 and y_2. We first show that κ is unique. For, from Gauss' equation the Riemann tensor of S is given by

$$^{(3)}R_{abcd} = \omega_{ac}\omega_{bd} - \omega_{ad}\omega_{bc} + R_{abcd}$$

in a tetrad adapted to the surface. But the tetrad rotation required to bring the space-like legs into the surface has a norm equal to $|N|$, which we have bounded by $3/2$, while bounds on R have been obtained in the last subsection. Hence on substituting these bounds we obtain an expression of the form

$$|^{(3)}R| < r_0 \left(K_1 x_0{}^2 r_0 + 1 \right)$$

for some constant K_1. Thus by choosing x_0 sufficiently small we can make $d(y_1, y_2)|^{(3)}R|$ sufficiently small, and so there will be a unique geodesic connecting y_1 to y_2 in S. Moreover, we can join each point of κ to x by a unique geodesic λ_t with $\lambda_t(0) = x$, $\lambda_t(1) = \kappa(t)$.

Set $\dot{\lambda}_t = K$, $\dot{\kappa}_t = H$. Then

$$\frac{d}{dt}(H.K) = (\nabla_H H).K + H.(\nabla_H K).$$

Since κ is a geodesic of S we have

$$\nabla_H H = N\omega(H, H)$$

and so

$$|K.\nabla_H H| < K_3 x_0{}^2 r_0 |H|^2$$

(using (39)). Moreover

$$H.\nabla_H K = H.\nabla_K H = H.H + H.Q \qquad (40)$$

where, from 2.1.3

$$|Q| < 1.32|H|r_0 k^2.$$

Now since κ is a geodesic with respect to the metric induced on S, which coincides with the metric on M for vectors tangent to S, we have that $H.H = \text{const.}$ Hence

$$\left| \frac{d}{dt}(H.K) - H.H \right| = \left| \frac{d}{dt}(H.K - tH.H|_0) \right|$$

$$< K_3 x_0{}^2 r_0 |H|^2 + |H|^2 K_2 r_0 x_0{}^2.$$

If we reparametrise κ so that $\kappa(1) = y_2$ and then integrate from y_1 to y_2 this then shows that

$$|H.H|_0 < K_4 |H|^2 x_0{}^2 r_0$$

for constants K_3 and K_4.

Clearly this becomes false for small enough $x_0{}^2 r_0$; in other words there is a value of x_0 such that for all $|X|$ smaller than this geodesics to S are unique.

Denote by $\xi(x)$ the endpoint on S of the unique maximal geodesic λ_x from x to S, which we parametrise by

$$\lambda_x(0) = x, \quad \lambda_x(1) = \xi(x) \in S.$$

We now define the coordinate z associated with this pair of p and q by

$$z(x) = (-\dot{\lambda}_x(1).\dot{\lambda}_x(1))^{1/2}. \tag{41}$$

Next we repeat the analysis of section 4.6.1 in order to calculate the smoothness of these coordinates, using as far as possible the same notation.

Draw an arbitrary curve $\mu : t \mapsto \mu(t)$, $\mu(0) = x$. Then corresponding to (24) we have

$$dz(\dot{\mu}(0)) = \frac{d}{dt} z(\mu(t))|_{t=0}. \tag{42}$$

Define $\bar{\lambda}$, X, Y as before, i.e.

$$\bar{\lambda}(t, s) = \lambda_{\mu(t)}(s), \quad X = \bar{\lambda}_*(\partial/\partial s), \quad \bar{Y} = \bar{\lambda}_*(\partial/\partial t),$$

and define $\mathbf{E}$ by choosing a basis at x, parallely propagating it along μ and then parallely propagating it along the curves λ_t. Thus (30),(26) and (29) still hold, while from (41) and (42) we have

$$dz(Y) = dz(\dot{\mu}(0)) = \frac{d}{dt}(-X.X)^{1/2} = -(1/z)X.\nabla_Y X$$

$$= -(1/z)X.\nabla_X Y = -(1/z)\frac{d}{ds}(X.Y). \tag{43}$$

But
$$\frac{d^2}{ds^2}(X.Y) = X_i \frac{d^2 Y^i}{ds^2} = X_i R^i{}_{jkl} Y^k X^j X^l = 0$$
i.e. $d(X.Y)/ds$ is constant. So we can write
$$(X.Y) = (X.Y)|_0 (1 - s)$$
(using the fact that X and Y are orthogonal at $s = 1$) giving
$$\frac{d}{ds}(X.Y) = -(X.Y)|_0.$$
Hence from (43)
$$dz(Y) = (1/z)(X.Y)$$
$$|dz| < (1/z)|X|$$
where the norm is evaluated in the frame defined above. So comparing this with (41) we see that $|dz|$ is bounded by the norm of N, which we have ensured is less than $3/2$. So, unlike normal coordinates, we have an absolute bound on dz independent of the Riemann tensor. This tensor appears at the next order of differentiation, where we have for any V in $T_x M$
$$(\nabla_Y dz)(V) = -\left(1/z^2\right)(Yz)(X.V) + (1/z)(\nabla_Y X.V)$$
$$= -\left(1/z^2\right)(X.Y)(X.V) + (1/z)(\nabla_Y X.V).$$
Hence from equation (32) we see that the first derivative of the metric in these coordinates is bounded by a constant multiplied by the norm of the Riemann tensor, provided that the size of the domain of the coordinates is less than a certain dimension, again determined by the Riemann tensor.

4.8 Smoothing

In the preceding section we have considered a smooth metric and found coordinates in which the derivatives of the metric can be estimated in terms of the Riemann tensor. We now show that the same construction works even if the metric is not as smooth as was assumed above. The method is to apply a smoothing operator to the metric.

We suppose throughout this section that the metric is given on a domain ω with compact closure in $\mathbb{R}^4$ (i.e. we are working in a coordinate patch); that the weak derivatives of g are locally bounded, so that the Riemann tensor is well defined as a distribution; and that the Riemann tensor components coincide with L^2 functions. $R^\alpha{}_{\beta\gamma\delta}$ will denote such a function.

Choose a non-negative C^∞ function $\rho : \mathbb{R} \to \mathbb{R}$ with compact support such that $\rho(x) = 1$ when $x < \epsilon$ for some $\epsilon > 0$, and

$$\int \rho(x)dx = 1$$

and define $\chi_n(x) = n^4\rho(n|x|)$ for $x \in \mathbb{R}^4$. Define a linear smoothing operator $S^{(n)}$ to be convolution with χ_n:

$$\left(S^{(n)}f\right)(x) = \int \chi_n(x - y)f(y)d^4y.$$

Let the metric $g^{(n)}$ be defined to have components $S^{(n)}g_{\alpha\beta}$, and denote all geometrical quantities defined from it by superscripted (n).

Proposition 4.8.1

$$R^{(n)\alpha}{}_{\beta\gamma\delta} \to R^\alpha{}_{\beta\gamma\delta} \quad \text{in } L^2(\omega).$$

Proof
See (Geroch and Traschen, 1987; Clarke 1982).

□

This is the key to applying all the previous results in cases where the final result is defined for lower differentiability, but where intermediate steps in the proof require smooth functions. The proposition shows that the operations of smoothing and differentiation, when applied to the production of the Riemann tensor from the metric, commute with each other. Therefore we can prove all our preceding results with lower differentiability by smoothing the metric, applying the result to the smoothed metric, and then letting $\epsilon \to 0$.

5

The analytic extension problem

5.1 Mathematical preliminaries

We shall be extending the metric on space-time by constructing integral expressions in which the metric coefficients are multiplied by a singular kernel. So our first task is to examine the properties of such expressions. The following theorem is a strengthening of a result in (Clarke, 1982).

Theorem 5.1.1

Suppose given a function ϕ with Hölder class C^α, viz

$$|\phi(\bar{x}) - \phi(x)| \leq K|\bar{x} - x|^\alpha, \qquad (1)$$

and define

$$v = J_1 + J_2 \qquad (2a)$$

$$J_1(x) = \int_{|x-z|<R} \phi(z)\theta(x, z)d^n z \qquad (2b)$$

$$J_2(x) = \int_{|x-z|>R} \phi(z)\theta(x, z)d^n z \qquad (2c)$$

where the Kernel θ has the form

$$\theta(x, z) = |x - z|^{-n+1}\hat{\theta}\left(x, \frac{z - x}{|z - x|}\right) \qquad for \ |z - x| < R, \quad (2d)$$

$\partial_i^x\theta$ is Hölder continuous in x for $|z - x| > R/2$, and $\hat{\theta}$, $\partial_i^x\hat{\theta}$ and $\partial_i^\omega\hat{\theta}$ are Hölder continuous in x, uniformly in ω.

Then the x-derivative of v satisfies a Hölder condition with exponent α.

Proof

Clearly it is sufficient to show this for J_1. Let η be a C^∞ function with $0 \leq \eta \leq 1$, $0 \leq \eta' \leq 2$, $\eta(t) = 0$ for $t \leq 1$ and $\eta(t) = 1$ for

96

$t \geq 2$. For $\epsilon > 0$ define

$$J_\epsilon(x) = \int_{|x-z|<R} \phi(z)\theta(x,z)\eta(|x-z|/\epsilon)d^n z$$

Then the derivative with respect to x is given by

$$\partial_i^x J_\epsilon(x) = -\int_{|x-z|<R} \phi(z) - \phi(x))\partial_i^z [\theta(x,z)\eta(|x-z|/\epsilon)] d^n z$$

$$- \phi(x)\int_{|x-z|<R} \partial_i^z [\theta(x,z)\eta(|x-z|/\epsilon)] d^n z +$$

$$+ \int_{|x-z|<R} \phi(z) [\partial_i^x \theta(x,z) + \partial_i^z \theta(x,z)] \eta(|x-z|/\epsilon)d^n z.$$

Performing a partial integration in the middle term and letting $\epsilon \to 0$ gives

$$\partial_i^x J_1(x) = J_3 + J_4 + J_5$$

where

$$J_3 = -\int_{|x-z|<R} \left(\phi(z) - \phi(x)\right) \partial_i^z [\theta(x,z)] d^n z$$

$$J_4 = -\phi(x)\int_{|x-z|=R} n_i\theta(x,z)d^n z$$

$$J_5 = +\int_{|x-z|<R} \phi(z) \left[\partial_i^x \hat{\theta}(x,z)\right] |x-z|^{-n+1} d^n z. \qquad (3)$$

The Hölder continuity of J_4 is easily seen, and so we must examine J_3 and J_5.

Set $\tilde{x} = (x + \bar{x})/2$ and set $d = |x - \bar{x}|$. Then

$$|J_5(\bar{x}) - J_5(x)| \leq I_1 + I_2 + 2I_3 + I_4$$

with

$$I_1 = \int_{d<|\tilde{x}-z|<R} \phi(z) \left[\partial_i^x \hat{\theta}(\bar{x},z) - \partial_i^x \hat{\theta}(x,z)\right] |\bar{x} - z|^{-n+1} d^n z$$

$$I_2 = \int_{d<|\tilde{x}-z|<R} \phi(z)\partial_i^x \hat{\theta}(\bar{x},z) \left[|\bar{x} - z|^{-n+1} - |x - z|^{-n+1}\right] d^n z$$

$$I_3 = +\int_{|\tilde{x}-z|<2d} \phi(z) \left[\partial_i^x \hat{\theta}(x,z)\right] |x - z|^{-n+1} d^n z$$

$$I_4 = +\int_{R-d<|\tilde{x}-z|<R+d} \phi(z) \left[\partial_i^x \hat{\theta}(x,z)\right] |x - z|^{-n+1} d^n z.$$

Inserting bounds and Hölder estimates for the θ-terms gives

$$|I_1| \leq \text{const.} \times d^\alpha; \quad |I_2| \leq \text{const.} \times d\log d; \quad |I_3|,|I_4| \leq d,$$

which is sufficient for a Hölder bound.

By boundedness it is sufficient to consider the case $d < R$. Then

$$|J_3(\bar{x}) - J_3(x)| \leq \bar{I}_5 + I_5 + I_6 + I_7$$

where

$$I_5 = \lim_{\epsilon \to 0} \int_\epsilon^{2d} \hat{\theta}_i(x,\omega) \left[\phi(x + r\omega) - \phi(x)\right] r^{-1} dr d^{n-1}\omega$$

$\bar{I}_5$ is the same with x replaced by $\bar{x}$,

$$I_6 = \int_{d<|\bar{x}-z|<R+d} (\phi(z) - \phi(\bar{x})) \left[\partial_i^z \theta(\bar{x}, z) - \partial_i^z \theta(x, z)\right] d^n z$$

$$I_7 = \int_{|\bar{x}-z|>R+d} [\phi(\bar{x})|\theta(\bar{x}, z) - \theta(x, z)|$$
$$+ |\phi(\bar{x})\theta(\bar{x}, z) - \phi(x)\theta(x, z)|] d^n z$$

Then

$$|I_5| \leq \lim_{\epsilon \to 0} CK \int_\epsilon^d r^{\alpha-1} dr = CK d^\alpha / \alpha$$

where K is the constant of (1) and $C = \sup_{x',\omega'} \hat{\theta}_i(x',\omega')\omega_{n-1}$ and ω_{n-1} is the area of the $(n-1)$-sphere.

We can write

$$\partial_z^i \theta(x, z) = |x - z|^{-n} \hat{\theta}_i(x,\omega),$$

where $\omega = (z - x)/|z - x|$, and

$$\hat{\theta}_i(x,\omega) = \partial_i^\omega \hat{\theta}(x,\omega) - (n-1)\omega^i \hat{\theta}(x,\omega)$$

(so that $\hat{\theta}_i$ is Hölder continuous in x:

$$|\hat{\theta}_i(\bar{x}) - \hat{\theta}_i(x)| \leq D d^\alpha)$$

Then

$$|I_6| = \left|\int [\phi(z) - \phi(x)] \left(\left(\hat{\theta}_i(\bar{x}, z) - \hat{\theta}_i(x, z)\right)|z - x|^{-n} + \right.\right.$$
$$\left.\left. + \hat{\theta}_i(x)\left(|z - x|^{-n} - |z - \bar{x}|^{-n}\right)\right) d^n z\right|$$

$$\leq \text{const.} \times \int_d^R r^\alpha \left(d^\alpha r^{-n} + dr^{-n-1}\right) r^{n-1} dr$$

$$\leq \text{const.} \times d^\alpha$$

as required. I_7 can be estimated in a similar way (though the task is somewhat easier) giving the final result that ∂v is Hölder continuous.

$\square$

5.1.1 The Calderón-Zygmund theorem

The result for Sobolev spaces corresponding to the previous one was proved by Calderón and Zygmund (1952, 1956). The special case of their work which we need is expressed by the following:

Theorem 5.1.2 *Suppose ϕ is a square-integrable function, $\theta(x, z)$ is given by (2d) with $\partial_i^x \hat{\theta}$ bounded and $\partial_i^x \theta$ square integrable in x for $|z - x| > R/2$, and v is given by (2a) - (2c). Then ∂_i^x is square integrable.*

Proof

As in the previous section, the derivative of v is given by (3). The L^2 norm of J_5 can be estimated from the bounds on $\partial_i^x \hat{\theta}$, while that of J_4 is immediate. The L^2 norm of J_3 follows from theorem 2 of Calderón and Zygmund (1956). $\qquad \square$

Corollary 5.1.3

If, in addition to the above conditions, $\partial_i^x \partial_j^x \hat{\theta}$ is bounded and $\phi \in H^s$, then $v \in H^{s+1}$.

Proof

If ϕ is differentiable then we can integrate by parts in (3b) to express the derivative of v in terms of the derivative of ϕ. A further differentiation with respect to x then allows us, from the preceding theorem, to estimate the L^2 norm of the second derivative of v in terms of the L^2 norm of the derivative of ϕ and the bounds on the second derivative of $\hat{\theta}$. The result then follows by interpolation (cf. 4.2.2) $\qquad \square$

5.2 Extensions of functions

5.2.1 The Calderon extension operator

Suppose we are given a C^{m+1} function f on a domain $\omega \subset \mathbb{R}^n$ with Lipshitz boundary: i.e. for any point x on the boundary there is a coordinate neighbourhood U of x in which the boundary is describable by

$$\partial \omega \cap U = \{x \mid x^0 < \psi(x^1, x^2, \ldots, x^{n-1})\}$$

with ψ a Lipshitz function –

$$|\psi(x) - \psi(y)| < K|x - y|$$

for a constant K. Suppose also that the derivatives of f are continuous on the closure of ω and hence tend to limits on the boundary of ω. Then we can construct an extension of f beyond ω to the whole of $\mathbb{R}^n$ by constructing a Taylor series for each point on the boundary using these limits, and then averaging over part of the boundary.

Explicitly, let

$$f^{(m)}(x, y) = \sum_{|\mathbf{k}| \leq m} f_{\mathbf{k}}(x)(y - x)^{\mathbf{k}} \frac{1}{\mathbf{k}!}$$

(the m'th-order Taylor series of f using derivatives at x and evaluated at y). We use a multi-index notation in which $\mathbf{k}, \mathbf{l}$, etc range over n-tuples of non-negative integers and we define

$$|\mathbf{k}| \equiv k^0 + \cdots + k^{n-1}, \qquad \mathbf{k}! \equiv k^1! \ldots k^{n-1}!$$

$$\partial_{\mathbf{k}} \equiv \frac{\partial^{|\mathbf{k}|}}{\partial x_0^{k^0} \ldots \partial x_{n-1}^{k^{n-1}}}, \qquad f_{\mathbf{k}} \equiv \partial_{\mathbf{k}} f,$$

$$x^{\mathbf{k}} \equiv \left(x^1\right)^{k^1} \ldots \left(x^{n-1}\right)^{k^{n-1}}$$

For $y \notin \omega$ set

$$\bar{f}(y) = \int \phi(\omega) f^{(m)}(y + \rho\omega, y) d^{n-1}\omega \tag{5}$$

where ω ranges over the unit $(n-1)$-sphere with the usual measure $d^{n-1}\omega$, ϕ is a smooth function on the unit sphere with unit integral and support in $\{\omega \,|\, \omega^0 < -K/\sqrt{(1 + K^2)}\}$ and $\rho = \rho(y, x)$ is such that $y + \omega\rho$ is on the boundary of ω in U (this ρ being unique for ω in the above support).

For some purposes it is useful to rewrite this extension as an integral over the whole of ω. Let χ be a C^∞ function, with compact support, equal to unity in $[0, 1]$, and suppose that y is close enough to the boundary so that $\rho < 1$ for all ω in the support of ϕ. Then an alternative to (5) is

$$\bar{f}(y) = \int \phi(\omega) \sum_{k \leq m} \chi(\rho) \frac{(-\rho)^k}{k!} \left[\left(\frac{d}{dr}\right)^k f(y + r\omega)\right]_{r=\rho}$$

which, by integration by parts, gives

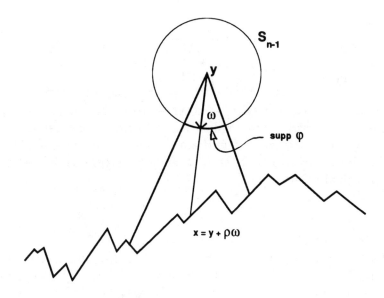

Fig. 1. Extension through a Lipshitz boundary

$$-\int \phi(\omega)\left[\int_\rho^\infty \chi(r)\frac{d}{dr}\left\{\sum_{k\le m}\frac{(-1)^k}{k!}\left(\frac{d}{dr}\right)^k f(y+r\omega)\right\}dr\right.$$

$$\left.+\int_\rho^\infty \chi'(r)\sum_{k\le m}\frac{(-1)^k}{k!}\left(\frac{d}{dr}\right)^k f(y+r\omega)dr\right]d^{n-1}\omega$$

Performing the r differentiation in the first integral transforms it into

$$-\int \phi(\omega)\left[\int_\rho^\infty \chi(r)\frac{(-1)^m}{m!}\left(\frac{d}{dr}\right)^{m+1} f(y+r\omega)dr\right]d^{n-1}\omega$$

$$=-\int \phi(\omega)\left[\int_\rho^\infty \chi(r)(-1)^m \sum_{|\mathbf{k}|=m+1} f_\mathbf{k}(y+r\omega)\omega^\mathbf{k}\frac{m+1}{k!}dr\right]\times$$

$$\times\, d^{n-1}\omega$$

Inserting this in (6) and integrating by parts k times in each term in the summation in the second integral thus gives

$$\bar{f}(y)=\int_\omega \sum_{|\mathbf{k}|=m+1} \lambda_\mathbf{k}(x-y)f_\mathbf{k}(x)d^n x + \int_\omega \mu(|x-y|)f(x)d^n x \quad (7)$$

where

$$\lambda_{\mathbf{k}}(z) = (-1)^{m+1}\phi(z/|z|)\chi(|z|)z^{\mathbf{k}}|z|^{-n}(m+1)/\mathbf{k}!$$

$$\mu(r) = -\sum_{k \le m} \left(\frac{d}{dr}\right)^k \frac{r^k}{k!}\chi'(r)$$

From (7), it is clear that, as x tends to the boundary of ω, $\bar{f}$ and its first m derivatives tend to f and its first m derivatives, so that $\bar{f}$ is a C^m extension of f. If $x \in \omega$, then the expression (7) is well-defined provided that the first singular integral is defined as the principal value

$$\lim_{\epsilon \to 0} \int_{\omega \setminus S_\epsilon^y} \tag{8}$$

S_ϵ^y being the ball of radius ϵ and centre y.

The forgoing argument, with the boundary of this sphere replacing the boundary of ω, shows that in this case $\bar{f}$ coincides with f. Thus (7) defines an extension for all x.

Next we compare the differentiability of this extension with that of the original function f in the case where we let f tend to a function that is not necessarily C^{m+1}.

5.2.2 Hölder norms

If f is a C^m function with a Hölder condition on the m'th derivatives given by $|f(x) - f(y)| < G|x - y|^\alpha$ then it is not hard to show from (7) that $\bar{f}$ is also of the same Hölder class. We can, however, do better than this by evaluating the quantity appropriate to a higher Hölder norm.

If we convert from an integral over ω to an integral over the boundary of ω using the Lebesgue measure corresponding to the coordinates x^1, x^2, x^3 then we can write

$$\bar{f}(x) = \sum_{|\mathbf{k}| \le m} \frac{1}{\mathbf{k}!} f_{,\mathbf{k}}(y_0)(x - y_0)^{\mathbf{k}} + \sum_{|\mathbf{k}| \le m} \int (f_{\mathbf{k}}(y)(x - y)^{\mathbf{k}}$$

$$- f_{\mathbf{k}}(y_0)(x - y_0)^{\mathbf{k}})\,\theta(x - y)\frac{1}{\mathbf{k}!}d^3y$$

where

$$\theta(z) = \phi\left(-\frac{z}{|z|}\right)\frac{z^0}{|z|^4}.$$

Hence for $|\mathbf{n}| > m$, differentiating and putting $\mathbf{v} = \mathbf{k} - \mathbf{l}$ gives

$$\partial_{\mathbf{n}} \bar{f}(x) = \sum_{\substack{\mathbf{l} \leq \mathbf{n} \\ |\mathbf{l}| \leq m}} \binom{\mathbf{n}}{\mathbf{l}} \int \sum_{|\mathbf{v}| \leq m - |\mathbf{l}|} (f_{\mathbf{v}+\mathbf{l}}(y)(x - y)^{\mathbf{v}}$$

$$- f_{\mathbf{v}+\mathbf{l}}(y_0)(x - y_0)^{\mathbf{v}}) \, \theta_{\mathbf{n}-\mathbf{l}}(x - y) \frac{1}{\mathbf{v}!} d^3 y \quad (9)$$

Now to estimate the term

$$Y \equiv \sum_{|\mathbf{v}| \leq m - |\mathbf{l}|} (f_{\mathbf{v}+\mathbf{l}}(y)(x - y)^{\mathbf{v}} - f_{\mathbf{v}+\mathbf{l}}(y_0)(x - y_0)^{\mathbf{v}}) \frac{1}{\mathbf{v}!}$$

we can choose for each y a point $y_1 \in \omega$ distant less than $K|y_0 - y|$ from both y and y_0, and express the f-terms as a Taylor expansion about y_1. We obtain

$$Y = \sum_{|\mathbf{v}| \leq m - |\mathbf{l}|} \frac{1}{\mathbf{v}!} \left[\sum_{|\mathbf{w}| \leq p} \frac{1}{\mathbf{w}!} f_{\mathbf{w}+\mathbf{v}+\mathbf{l}}(y_1) \left\{ (x - y)^{\mathbf{v}}(y - y_1)^{\mathbf{w}} \right. \right.$$

$$- (x - y_0)^{\mathbf{v}}(y_0 - y_1)^{\mathbf{w}} \}$$

$$+ \sum_{|\mathbf{w}| \leq p} \frac{1}{\mathbf{w}!} \{ f_{\mathbf{w}+\mathbf{v}+\mathbf{l}}(\hat{y})(x - y)^{\mathbf{v}}(y - y_1)^{\mathbf{w}}$$

$$\left. - f_{\mathbf{w}+\mathbf{v}+\mathbf{l}}(\hat{y}_0)(x - y_0)^{\mathbf{v}}(y_0 - y_1)^{\mathbf{w}} \} \right]$$

where $p = m - |\mathbf{v}| - |\mathbf{l}| - 1$ and where $\hat{y}$ (resp. $\hat{y}_0$) is a point on the straight coordinate line between y_1 and y (resp y_0). Putting $\mathbf{z} = \mathbf{v} + \mathbf{w}$ we can rewrite the above as

$$Y = \sum_{|\mathbf{z}| \leq p'} \sum_{|\mathbf{w}| \leq |\mathbf{z}|} \frac{1}{\mathbf{w}!(\mathbf{z} - \mathbf{w})!} f_{\mathbf{z}+\mathbf{l}}(y_1) \left\{ (x - y)^{\mathbf{z}-\mathbf{w}}(y - y_1)^{\mathbf{w}} \right.$$

$$- (x - y_0)^{\mathbf{z}-\mathbf{w}}(y_0 - y_1)^{\mathbf{w}} \}$$

$$+ \sum_{|\mathbf{w}|=p''} \sum_{|\mathbf{w}| \leq |\mathbf{z}|} \frac{1}{(\mathbf{z} - \mathbf{w})!\mathbf{w}!} \{ f_{\mathbf{z}+\mathbf{l}}(\hat{y})(x - y)^{\mathbf{z}-\mathbf{w}}(y - y_1)^{\mathbf{w}}$$

$$- f_{\mathbf{z}+\mathbf{l}}(\hat{y}_0)(x - y_0)^{\mathbf{z}-\mathbf{w}}(y_0 - y_1)^{\mathbf{w}} \}$$

where $p' = m - |\mathbf{l}| - 1$, $p'' = m - |\mathbf{l}|$, i.e.

$$Y = \sum_{|\mathbf{z}|=p''} [f_{\mathbf{z}+\mathbf{l}}(\hat{y}) - f_{\mathbf{z}+\mathbf{l}}(\hat{y}_0)] \frac{(x - y_1)^{\mathbf{z}}}{\mathbf{z}!}$$

from the binomial theorem. Substituting in (9) thus gives

$$\partial_{\mathbf{n}} \bar{f}(x) = \sum_{\substack{1 \leq \mathbf{n} \\ |\mathbf{l}| \leq m}} \binom{\mathbf{n}}{\mathbf{l}} \int \sum_{|z| \leq m - |\mathbf{l}|} (f_{z+1}(\hat{y}) - f_{z+1}(\hat{y}_0))\, \theta_{\mathbf{n}-\mathbf{l}}(x-y) \times$$

$$\times \frac{(x - y_1)^z}{z!} d^3 y.$$

The difference of the two fs can now be estimated by the Hölder norm on f, while the derivative of θ is of the order of magnitude of $h^{-|\mathbf{n}|+|\mathbf{l}|-3}$ where h is the x^0-coordinate distance of x from the boundary of ω. Estimating all the terms as functions of h and performing the integration then gives

$$|\partial_{\mathbf{n}} \bar{f}(x)| \leq B_{\mathbf{n}} h^{\alpha + m - |\mathbf{n}|} \tag{10}$$

for some constant $B_{\mathbf{n}}$.

5.2.3 Sobolev norms

We now consider the effect of the extension operator

$$E_m : f \to \bar{f}$$

defined by (7) in terms of the Sobolev norms

$$(\|f\|_v)^2 = \int_\omega \sum_{|\mathbf{k}| \leq v} |\partial_{\mathbf{k}} f|^2 d^n x$$

interpolated to non-integral v.

It can be shown that, for $m \leq v \leq m + 1$,

$$\|\bar{f}\|_{v,2} \leq C \|f\|_{v,2} \tag{11}$$

for some constant C depending on v and ω; we shall outline a proof given (for a slightly different operator) in more detail in (Adams, 1975, p. 91).

Using (7) we have that

$$\partial_{\mathbf{l}} \bar{f}(y) = \int_\omega \sum_{|\mathbf{k}|=m+1} \lambda_{\mathbf{k},\mathbf{l}}(x-y) f_{\mathbf{k}}(x) d^n x + \int_\omega \mu_{,\mathbf{l}}(|x-y|) f(x)^n dx \tag{12}$$

where the first integral is defined in the sense of (8).

Now $\mu_{,\mathbf{l}}(|x - y|)$ is a C_0^∞ function (because $\chi'(r)$ is zero in a neighbourhood of $r = 0$) and so the second integral can immediately be estimated in terms of $\|f\|_0$. Also $\lambda_{\mathbf{k}}(z)$ is a function homogeneous of order $m + 1 - n$ in $|z| < 1$, and is C_0^∞ outside this region. Thus when $|\mathbf{l}| \leq m$, we see that $\lambda_{\mathbf{k},\mathbf{l}}$ is of homogeneity degree greater than or equal to $1 - n$ and thus is in L^1. In this case Young's theorem (Adams, 1975) shows that for any $g \in L^2(\omega)$ we have

$$\|\lambda_{\mathbf{k},\mathbf{l}} * g\|_2 \leq \|\lambda_{\mathbf{k},\mathbf{l}}\|_1 \|g\|_2$$

(where $*$ denotes convolution), enabling us to bound the first integral as required.

For $|\mathbf{l}| = m + 1$ a more powerful result is needed. In this case $\lambda_{\mathbf{k},\mathbf{l}}(z)$ is homogeneous of degree $-n$ in $|z| < 1$ and satisfies

$$\int_{|z|=1} \lambda_{\mathbf{k},\mathbf{l}}(z) d^{n-1}z = 0.$$

For such functions the result of (Calderon and Zygmund, 1952, 1956) (cf. 5.1.2) enables one to bound $\|\lambda_{\mathbf{k},\mathbf{l}} * g\|_2$ by a multiple of $\|g\|_2$, as required. Thus we have established (11) for $v = m + 1$.

Next we show that this result extends to non-integral $v = m$. For, suppose $|\mathbf{l}| = m$ in (12) and consider one of the terms in the summation in the first integral. For some i we will have $k^i \neq 0$. Integrate by parts on the ith variable to give

$$\int_\omega \sum_{|\mathbf{k}|=m+1} \lambda_{\mathbf{k},\mathbf{l},i}(x - y) f_{\mathbf{k}'}(x) d^n x$$

where $\mathbf{k}'$ is $\mathbf{k}$ with the i'th coordinate reduced by 1. The Calderon-Zygmund theory then allows us to estimate this in terms of $\|f\|_m$ proving (11) for $v = m$. Finally, the interpolation result 4.2.2 gives the formula for general v.

5.2.4 Action of singular kernel

For later use we will need to know the effect of integrating an extended function with a singular kernel of the form considered in 5.1.1. In the Sobolev case there is no problem since both the extension and the singular integral are, as we have seen, covered by the Calderón-Zygmund theorem. But in the Hölder case we need to show that equation (10) is preserved in an appropriate sense.

Theorem 5.2.1 *Let $\bar{f}$ be the extension of f, satisfying (10) and define v as in equation (2) with ϕ replaced by $\bar{f}$ and the same* *conditions on θ. Then the first partial derivatives of v satisfy (10) (with a different constant): i.e.*

$$|\partial_{\mathbf{n}} v| \leq C_{\mathbf{n}} h^{\alpha+m-|\mathbf{n}|+1}$$

for $|\mathbf{n}| \geq 2$.

We give the proof only for the case that will be needed; the more general result is obtained by partially integrating to reduce to this case.

Proof

(Case $m = 0$, $|\mathbf{n}| = 2$).

We evaluate the second derivative of v in the same way as in equation (3). As there, we need only verify the result for the part of $|\partial_{\mathbf{n}} v|$ given by

$$J(x) = \int_{|x-z|<R} \bar{f}(z)\theta(x,z)d^n z$$

which we approximate by

$$J_\epsilon(x) = \int_{|x-z|<R} \bar{f}(z)\theta(x,z)\eta(|x-z|/\epsilon)d^n z$$

Then

$$\partial_j^x \partial_i^x J_\epsilon = \int \bar{f}(z)(\partial_j^x + \partial_j^z)(\partial_i^x + \partial_i^z)(\theta\eta)d^n z$$
$$- 2\int \bar{f}(z)\partial_{(j}^z(\partial_{i)}^x + \partial_{i)}^z)(\theta\eta)d^n z + \int \bar{f}(z)\partial_j^z\partial_i^z(\theta\eta)d^n z$$

the domain of integration being $U_R := |x - z| < R$.

If we split the range of integration into $U_1 = U_R \cap \omega$ and $U_2 = U_R \backslash \omega$, then the part of the above expression arising from U_2 can be written, by partial integration and using $(\partial^x + \partial^z)\eta = 0$, as

$$\int_{U_2} \bar{f}(z)\left[(\partial_j^x + \partial_j^z)(\partial_i^x + \partial_i^z)\theta\right]\eta d^n z$$
$$+ 2\int_{U_2} (\partial_{(j}^z \bar{f}(z))(\partial_{i)}^x\theta + \partial_{i)}^z\theta)\eta d^n z$$
$$- 2\int_{\partial U_2} \bar{f}(z)(\partial_{(i}^x\theta + \partial_{(i}^z\theta)\eta d^{n-1}S_{j)}$$
$$- \int_{U_2} \left(\partial_j^z\left(\bar{f}(z) - (z-x)^k\bar{f}_{,k}(x)\right)\right)\partial_i^z\theta d^n z$$
$$+ \int_{\partial U_2} \left(\left(\bar{f}(z) - (z-x)^k\bar{f}_{,k}(x)\right)\right)\partial_i^z\theta d^{n-1}S_j$$

$$+ (z - x)^k \bar{f}_{,k}(x) \int_{\partial U_2} \partial_i^z (\theta \eta) d^{n-1} S_j.$$

A direct evaluation now gives that all these terms are bounded either by const $\times h$ or const $\times h^{\alpha-1}$, as required.

For the contributions from U_1 , the only term that cannot immediately be bounded is

$$\int_{U_1} \bar{f}(z) \partial_j^z \partial_i^z \theta d^n z$$

(noting that for $x \in \omega$ we may take the limit $\epsilon \to 0$ at once), which we rewrite as

$$\int_{U_1} (\bar{f}(z) - \bar{f}(x^*)) \partial_j^z \partial_i^z \theta d^n z + \bar{f}(x^*) \int_{\partial U_1} \partial_i^z \theta d^{n-1} S_j$$

for x^* , a function of x, chosen in ω distant less than $2h$ from x. The Hölder norm on $\bar{f}$ now allows us to show that the first term is bounded by $h^{\alpha-1}$, while the second term is bounded. This establishes the theorem. $\qquad \Box$

5.3 The Poincaré lemma

This says that if ϕ is a p-form satisfying $d\phi = 0$ then there exists χ such that $\phi = d\chi$. The usual proof of this involves the construction of an operator I satisfying

$$d(I\phi) = Id\phi + \phi \tag{13}$$

for all forms ϕ. The lemma then follows from taking $\chi = I\phi$. Thus if $\bigwedge^p$ denotes the space of p-forms on the domain in question, I maps $\bigwedge^p$ to $\bigwedge^{p+1}$ for all $p > 0$. Our aim here is to show that I can be defined so as to increase the level of differentiability by 1. The domain ω in $\mathbb{R}^n$ on which the forms are defined has to fairly regular, and for simplicity we suppose it to be star-shaped from the origin and of diameter less than R.

Theorem 5.3.1

Under the above conditions on ω, there exists an operator I satisfying (13) such that, if the components of ϕ are bounded in a Hölder (resp. a Sobolev) norm, then the components of $I\phi$ are bounded in the norm with one higher level of differentiability.

Proof

Define $I_0(x_0) : \bigwedge^p \to \bigwedge^{p+1}$ by

$$[I_0(x)\phi](y) = \sum_{\mu_1 < ... < \mu_p} \sum_{i=1}^{p} (-1)^{i-1} \int_0^1 t^{p-1} \phi_{\mu_1 ... \mu_p}(x_0 + t[y - x_0])$$
$$(y^{\mu_i} - x_0^{\mu_i}) dt \, dx^{\mu_1} \wedge ... \wedge dx^{\mu_{i-1}} \wedge dx^{\mu_{i+1}} \wedge ... \wedge dx^{\mu_p}$$

$$(14)$$

where ϕ is specified by

$$\phi = \sum_{\mu_1 < ... < \mu_p} \phi_{\mu_1 ... \mu_p} dx^{\mu_1} \wedge ... \wedge dx^{\mu_p}.$$

It is then well known (see, for example, Spivak (1965) pp. 94-95) that

$$dI_0(x)\phi = \phi - I_0(x)d\phi.$$

(The condition that ω is star-shaped enters in order to ensure that the integrand in the above definition is well-defined for x_0 near enough to the origin.)

I is then defined by "smearing" $I_0(x)$ with respect to x. Take a C^∞ non-negative function ξ with support in a ball of radius 1, centre the origin, in $\mathbb{R}^n$ and satisfying $\int \xi(x)dx = 1$. Choose a so small that ω is star-shaped from every point in a ball of radius a centre the origin, and set

$$(I\phi)(y) = a^{-(n+1)} \int \xi\left(\frac{x}{a}\right) [I_0(x)\phi](y)d^n x. \tag{15}$$

This will still satisfy (13). If we substitute the definition (14) and change variables to z, u with

$$x = y + u(z - y), \qquad t = 1 - 1/u$$

then (15) becomes

$$(I\phi)_{\mu_1 ... \mu_{p-1}}(y) = p \int \theta^\nu(y, z)\phi_{\nu\mu_1 ... \mu_{p-1}}(z)d^n z$$

with

$$\theta^\nu(y, z) = a^{-(n+1)} \int_{u_1}^{u_2} u^{n-2} \left(1 - \frac{1}{u}\right)^{p-2} (y^\nu - z^\nu) \times$$
$$\times \xi([y + u(z - y)]/a)du$$

A direct conversion of the limits on t in (14) will give $u_1 = 1$, $u_2 = \infty$; but the integrand will only be non-zero between

$$u_1 = \max(1, [|y| - a]/|z - y|)$$

and
$$u_2 = [|y| + a]/|z - y|,$$
which can therefore be taken as the limits of integration.

Calculation then shows that
$$|\theta^\nu(y, y + w) - |w|^{-n+1}\theta^\nu(y, \omega)| \leq \text{cst.} \times |w|^{-n+2}$$
where
$$\omega = w/|w| \quad \text{and} \quad \hat{\theta}^\nu(y, \omega) = \lim_{\lambda \to 0} \lambda^{-n+1}\theta^\nu(y, y + \lambda\omega)$$

The part $|w|^{-n+1}\hat{\theta}^\nu$ then satisfies the conditions of 5.2.3, 5.2.2, while the difference between this and θ^ν is of lower order, and so the result follows. $\qquad\qquad\square$

Note that this operator I does not preserve boundedness conditions, in the sense that if the components of ϕ are known to be bounded, there is no guarantee that the derivatives of $I\phi$ will be bounded. It is this that prevents us making use of this operator when it comes to Sobolev classes of space-times.

5.4 Extension of geometry

In the next chapter we shall find coordinates near certain boundary points in which a neighbourhood of the point is represented by a domain ω in $\mathbb{R}^4$, which we may take to be star-shaped with a locally Lipshitz boundary, on which the geometry is reasonably well-behaved. We now sudy the problem of extending this geometry (the metric and its curvature) to a neighbourhood of ω. Thus we are ignoring the possibility of finding enlarged solutions to any field equations that may be satisfied on ω.

In the Sobolev case we shall require directly that the components of the metric are in H^s, so that extension is immediate using the extension operator. But in the Hölder case we can work with the Riemann tensor as the primary field, rather than the metric. In this case, in order to extend the geometry, two courses can be followed: in the first, one extends the metric components, with the problem that the resulting curvature may become highly singular. In the second, one extends the components of the curvature, with the problem that the resulting tensor may not be the curvature of any metric. The first problem is circumvented, of course, if, given that we are working at the level of C^k curvature, we can find co-ordinates in which the metric is C^{k+2}; but unfortunately there is no guarantee that this can be done. As we saw in 4.3, the best we can hope for is coordinates in which the metric is C^{k+1}. We follow the first course, extending (part of) the curvature while ignoring, at first, its relation to a metric. Then we adjust the geometry so that the curvature and the metric "fit". The "adjustment" involves constructing operators that would reconstruct the metric from the curvature, if the curvature were to come from a metric. The main step in this is the reconstruction of the connection from (part of) the curvature.

5.4.1 Extension of the connection

It is the highest-derivative terms of the curvature that are liable to be worst behaved, so we concentrate on these, writing

$$\mathbf{R}^\alpha_\beta := \left(\frac{\partial \Gamma^\alpha_{\beta\delta}}{\partial x^\gamma} \right) dx^\gamma \wedge dx^\delta$$

for the part of the curvature form (in a coordinate basis) corresponding to the highest derivatives. Note that if the Riemann tensor is C^k, then so are the components of the 2-form $\mathbf{R}^\alpha_\beta$. (For simplicity of expression, we shall give differentiability simply in terms of C^k-classes in this section; but it is to be understood that the corresponding statements for Sobolev and Hölder classes hold as well.)

We assume that the metric, and hence $\mathbf{R}$, are initially defined on a star-shaped domain ω and that $\mathbf{R}$ is then extended by the results of the first section to a larger such domain V. By the linearity of the extension operator, it is clear that the algebraic symmetries of $\mathbf{R}$ are preserved by this extension, but not, of course, the relation to the metric which is not yet defined in the larger region. We then proceed to construct a connection related to the extended curvature.

The relationship between the connection and the curvature is given by

$$\mathbf{R}^\alpha_\beta = d\Gamma^\alpha_\beta \tag{16}$$

where

$$\Gamma^\alpha_\beta := \Gamma^\alpha_{\beta\gamma} dx^\gamma.$$

In consequence of (16) we have

$$d\mathbf{R}^\alpha_\beta = 0. \tag{17}$$

So to reconstruct Γ we must solve (16) subject to the zero-torsion condition, which in this notation becomes

$$\Gamma^\alpha_\beta \wedge dx^\beta = 0. \tag{18}$$

If we apply the Poincaré lemma (5.3) to (16), with the integrability condition (17), then we see that $\bar{\Gamma}^\alpha_\beta = (I\mathbf{R})^\alpha_\beta$ is a solution to (16). But this solution is not unique: we have the freedom to add any exterior derivative, which we use to find a solution satisfying the subsidiary condition (18). As a first approximation to such a solution, we modify $I\mathbf{R}$ by the addition of an exterior derivative, to

$$I\mathbf{R}^\alpha_\beta - d\left(\left\{I\left([I\mathbf{R}^\alpha_\gamma] \wedge dx^\gamma\right)\right\}_\beta\right) = \gamma^\alpha_\beta \tag{19}$$

say, where $\{\quad\}_\beta$ denotes the β-th coordinate-component of the 1-form contained in the $\{\quad\}$. Counting levels of differentiability, we note that γ^α_β is C^{k+1}.

Then

$$\gamma^\alpha_\beta \wedge dx^\beta = \left(I\mathbf{R}^\alpha_\beta\right) \wedge dx^\beta - d\left\{I\left([I\mathbf{R}^\alpha_\gamma] \wedge dx^\gamma\right)\right\}$$

$$= Id\left([I\mathbf{R}^\alpha_\gamma] \wedge dx^\gamma\right)$$

(from (13))

$$= I\left(\left\{d\left[I\mathbf{R}^\alpha_\beta\right]\right\} \wedge dx^\gamma\right)$$

$$= I\left(\mathbf{R}^\alpha_\beta \wedge dx^\gamma\right)$$

(from (13), (17))

$$= 0$$

(from the Riemann symmetries.)

Thus γ, like Γ, satisfies both (18) and (16). So defining their difference as

$$\chi^\alpha_\beta := \gamma^\alpha_\beta - \Gamma^\alpha_\beta$$

(so that the components of χ^α_β are C^k, like those of Γ) we have

$$d\chi^\alpha_\beta = 0, \tag{20}$$

$$\chi_\beta^\alpha \wedge dx^\beta = 0. \tag{21}$$

From (20) we have

$$\chi_\beta^\alpha = d\sigma_\beta^\alpha \tag{22}$$

$$= \sigma_{\beta,\gamma}^\alpha dx^\gamma \tag{23}$$

where we can take

$$\sigma_\beta^\alpha = I\chi_\beta^\alpha,$$

making σ_β^α C^{k+1}, while (21) and (23) give

$$\sigma_{[\beta,\gamma]}^\alpha = 0,$$

i.e.

$$\sigma_\beta^\alpha = \omega_{,\beta}^\alpha$$

where

$$\omega^\alpha = I\left(\sigma_\beta^\alpha dx^\beta\right),$$

so that ω^α is C^{k+2}.

To summarise, the connection on ω is given by

$$\Gamma_{\beta\gamma}^\alpha = (\gamma_\beta^\alpha)_\gamma - \omega_{,\beta\gamma}^\alpha \tag{24}$$

where γ_β^α is given by (19) and the ω^α are C^{k+2} functions.

Thus if we extend $\mathbf{R}_\beta^\alpha$ and ω^α to V and then define $\Gamma_{\beta\gamma}^\alpha$ by (19) and (24), we obtain a torsion-free C^k connection which extends the original connection, and which, by construction, satisfies (16) and hence gives rise to a C^k curvature.

5.4.2 Extension of the metric

We now assume that the Riemann tensor and the connection on ω satisfies a Hölder condition of exponent α. In addition we require that the boundary of ω be Lipshitz. (Similar results to those in this subsection could be proved assuming Hölder conditions on derivatives of the Riemann tensor.)

We have seen that this will be preserved in the extension, and that $\mathbf{R}$ also satisfies the condition (10) with $m = 0$, in this case. From 5.2.4 and the definition of γ we see that the full Riemann tensor also satisfies (10). Then the extended metric is defined by parallelly propagating the metric in ω, using Γ, along straight coordinate lines from the origin.

To calculate the metric and its derivatives at a point $x_0 \notin \omega$, choose vectors $\underset{i}{E}$ at x_0 with $\underset{i}{E^\lambda} = \delta_i^\lambda$ and parallely propagate them first along the curve

$$u \to x_0^\lambda + u\delta_\nu^\lambda$$

then along the curves

$$t \to x_0^\lambda + u\delta_\nu^\lambda + t\delta_\mu^\lambda$$

for each fixed u and finally along the lines through the origin given by

$$s \to s(x_0^\lambda + u\delta_\nu^\lambda + t\delta_\mu^\lambda) =: \xi(s,t,u)^\lambda$$

for each fixed t, u. Setting

$$X^\lambda = \frac{\partial \xi^\lambda}{\partial s},$$

$$Y^\lambda = \frac{\partial \xi^\lambda}{\partial t} = s\delta_\mu^\lambda$$

$$Z^\lambda = \frac{\partial \xi^\lambda}{\partial u} \tag{25}$$

gives

$$\frac{D}{Ds} \underset{i}{E} = \nabla_X \underset{i}{E} = 0 \tag{26}$$

and hence

$$\frac{d}{ds} \underset{j}{E} . \nabla_Z \underset{i}{E} = \underset{j}{E} . \nabla_X \nabla_Z \underset{i}{E} = \underset{j}{E} . R(X,Z) \underset{i}{E}$$

which gives

$$\underset{j}{E} . \nabla_Z \underset{i}{E} \bigg|_{\xi(s,0,u)} = \int_s^1 \left[\underset{j}{E} . R(X,Z) \underset{i}{E} \right]_{(x=s'(x_0+u\delta_\nu))} ds'. \tag{27}$$

We then define the extended metric by requiring

$$\nabla_X g = 0 \tag{28}$$

so that

$$\nabla_X \left[(\nabla_Y g) \left(\underset{i}{E}, \underset{j}{E} \right) \right] = \frac{d}{ds} \left(s g_{\alpha\beta;\mu} \underset{i}{E^\alpha} \underset{j}{E^\beta} \right)$$

from (25)

$$= (\nabla_X \nabla_Y g) \left(\underset{i}{E}, \underset{j}{E} \right) = -2 \left[R(X,Y) \underset{(i}{E} \right] . \underset{j)}{E}$$

Integrating from ρ to 1 (where $\rho = \rho(x_0) = \sup\{s \,:\, \xi(s,0,0) \in \omega\}$) gives, from (28),

$$g_{\alpha\beta;\mu} = -2 \int_\rho^1 R^\rho_{\gamma\sigma\delta} Y^\delta x^\sigma \underset{(\alpha\ \ \beta)}{E^\gamma E_\rho} ds \tag{28'}$$

where we can take the left hand side as evaluated at $\xi(1,0,u)$, in which case the integrand is evaluated at $\xi(s',0,u)$.

A further differentiation yields

$$g_{\alpha\beta;\mu\nu} = (\nabla_Z \nabla_Y g)\left(\underset{\alpha}{E},\underset{\beta}{E}\right) = \nabla_Z (g_{\alpha\beta;\mu})$$

$$= -2 \int_\rho^1 R^\rho_{\gamma\sigma\delta;\tau} Y^\delta x^\sigma \left(\underset{(\alpha\ \ \beta)}{E^\gamma E_\rho}\right) Z^\tau ds$$

$$-2 \int_\rho^1 R^\rho_{\gamma\sigma\delta} \left(Y^\delta x^\sigma \underset{(\alpha\ \ \beta)}{E^\gamma E_\rho}\right)_{;\tau} Z^\tau ds \tag{29}$$

The derivative of R appearing in the first term is integrable (as a function of s), from 5.2.4, while derivatives of E in the second term can be estimated from (27). Moreover, the estimates obtained in this way can be extended to estimates of the Hölder constant for $g_{\alpha\beta;\mu\nu}$ by a direct estimation of $|g_{\alpha\beta;\mu\nu}(x) - g_{\alpha\beta;\mu\nu}(y)|$ from (29) (for details see Clarke, 1982, p. 302). The result is that

$$|g_{\alpha\beta;\mu\nu}(x) - g_{\alpha\beta;\mu\nu}(y)| < \text{const} \times |x - y|^\sigma$$

with the constant given in terms of the Hölder norms of R and the connection.

Our aim is to estimate the Riemann tensor of the connection derived from g. There is no problem with the terms quadratic in Γ since we know that g has Hölder continuous derivatives. The terms linear in Γ depend on the second derivatives of g through $g_{[\alpha[\beta,\gamma]\delta]}$. But this differs from $g_{[\alpha[\beta;\gamma]\delta]}$ only by terms in Γ; thus the above estimate of the Hölder continuity of $g_{[\alpha[\beta;\gamma]\delta]}$ gives us the Hölder continuity of the Riemann tensor of g.

6

Attributes of singularities

In this chapter I shall describe some of the concepts used to classify singularities, starting with properties related to differentiability and then dealing with those of a "topological" nature.

6.1 Strengths of singularities

6.1.1 Extension strengths

In section 3.4 we defined a singularity as a b-boundary point that cannot be removed by making an extension of the space-time. We can subdivide this notion according to the differentiability classes introduced in section 4.2. Let "P" denote a differentiability-class of space-time. I.e. "P" is "C^X" or "$\mathbf{C}^X$" (cf. 4.2.1) with "X" one of "∞","k", "0","k, α","$0-$" or "$k-$"; or "P" is $\mathbf{H}^s$ or H^s (cf 4.5). Then a singularity p is said to be a P-singularity if it is the endpoint of an incomplete curve γ and there is no extension θ of M of type P in which $\theta \circ \gamma$ is extendible.

One might ask whether there were singularities that were so bad that there could be no extension of the space-time manifold at all, irrespective of what conditions were put on the metric. But such "absolute singularities" cannot exist. For, suppose $\gamma : [0, 1) \to M$ is an incomplete curve as in the above definition ending at a singularity p. We can, by making arbitrarily small deformations, assume that γ is embedded in M and so we can find a neighbourhood U of γ diffeomorphic to $\mathbb{R}^4$ by a coordinate map $x : U \to \mathbb{R}^4$ with

1. $x(U) = \{\xi \mid -1 < \xi^0 < 0, (\xi^1)^2 + (\xi^2)^2 + (\xi^3)^2 < 1\}$
2. $x(\gamma(s))^0 = s$
3. for any sequence $\{p_n\}$ in U, $x(p_n)^0 \to 1 \Rightarrow p_n \to p$.

Then the manifold can be extended by extending $x(U)$ to the larger set $\{\xi \mid -1 < \xi^0 < 1, (\xi^1)^2 + (\xi^2)^2 + (\xi^3)^2 < 1\}$ and using the coordinate x to "glue" this patch onto M. Thus every singularity admits an extension provided that the metric is allowed to become singular.

A C^{0-} singularity is one where there is no extension in which the metric and its inverse are bounded. This appears to be the case for the Friedmann big-bang singularity, though even this is not obvious because every regular space-time, including Friemann, possesses an atlas (defined by taking normal coordinates in a small enough neighbourhood of each point – cf. 4.6.1) in which the components of the metric and its inverse are bounded uniformly on M; but since the images of the normal coordinate charts in $\mathbb{R}^4$ become progressively smaller as one approaches the singularity, these charts cannot be extended to cover the singularity itself, in any sense, and so cannot be used to form an extension of M. Since the Riemann tensor governs the size (in $\mathbb{R}^4$) of normal coordinate neighbourhoods with a given bound on the metric, it seems likely that, in those cosmologies where there are singularities for which the Riemann tensor remains bounded in a suitably Lorentz-boosted frame (defined as "p.p. singularities"), there are extensions in which the metric remains bounded.

A $\mathbf{C}^{0-}$ singularity is one where the Riemann tensor is unbounded in a neighbourhood of some point in any extension, which implies that this is also a C^{2-} singularity. An $H^{2.5+\epsilon}$ (or $\mathbf{H}^{0.5+\epsilon}$) singularity is one where no extension is possible with the metric having the differentiability implied by the existence theorems for Einstein's equations. This is probably the best candidate for a "real" singularity.

6.1.2 Limiting strengths

The problem with the above type of definition is that it is highly non-constructive: strength is defined in terms of the non-existence of an extension, rather than in terms of directly measurable quantities. A more useful concept of strength depends on examining the limiting behaviour of the curvature as one approaches the singularity.

The simplest course here is to work with the components of the covariant derivatives $R_{ijkl;m\cdots p}$ of the Riemann tensor in a given frame; these can be regarded as functions on the frame bundle. A function f on the frame bundle LM can be described in terms of Hölder classes as follows: we say that f is $C^{0,\alpha}$ (respectively C^{0-}, C^{1-}) at $p \in \partial_b M$ if for any $u \in \bar{\pi}^{-1}(p)$ there exists a neighbourhood U of u in $\overline{LM}$ and a constant K such that, for all x, y in $U \cap LM$,

$$|f(x) - f(y)| < K d(x, y)^\alpha$$

(respectively,

$$|f(x)| < K,$$
$$|f(x) - f(y)| < K d(x, y)$$

).

Similarly we can say that f is C^0 at p if, for any $u \in \bar{\pi}^{-1}(p)$ there exists a constant z such that $f(x) \to z$ as $x \to u$.

Then we say that p is a $\mathbf{C}^X$ *curvature singularity* if the components of the Riemann tensor or its kth covariant derivatives at p satisfy:

$\mathbf{C}^{k,\alpha}$ (resp. $\mathbf{C}^k$)   : $R_{ijkl;m\cdots p}$ (k derivs.) are not C^α (resp. C^0)
$\mathbf{C}^{k-}$: $R_{ijkl;m\cdots p}$ ($k - 1$ derivs) are not C^{1-}
$\mathbf{C}^{0,\alpha}$ (resp. $\mathbf{C}^{0-}$, $\mathbf{C}^0$)   : R_{ijkl} are not C^α (resp. C^{0-}, C^0).

If a singularity p is not a $\mathbf{C}^{0-}$ curvature singularity, i.e. if the Riemann tensor components are bounded near any point in the fibre over p, then we call p a *quasi-regular* singularity. It will turn out that quasi-regular singularities are associated with peculiarities in the topology of space-time; or, to put it another way, if the Riemann tensor is bounded near a boundary point and the space-time is topologically well-behaved there, then that boundary point is not a C^1-singularity at all – there is an extension. The later sections of this chapter will make precise the idea of "topologically well-behaved".

The main property of quasi-regular singularities is expressed by the following result that will be used later on. To formulate it, consider a timelike curve $\gamma : [0, 1) \to M$. By a *family generated by γ up to a* we mean a family of geodesics $\{\kappa_s \mid s \in [0, a)\}$ having, for all valid s, $\kappa_s(0) = \gamma(0)$ and $\kappa_s(1) = \gamma(s)$. When $a = 1$ we shall speak simply of a family generated by γ.

Proposition 6.1.1

Suppose $p \in \overline{LM}$ is such that $R^i{}_{jkl}$ is bounded in a neighbour-

hood of p. Then there exists an $\epsilon > 0$ such that, for any horizontal curve of length $< \epsilon$ ending at p with future timelike projection γ in M, either there is a congruence generated by γ up to a with $a < 1$ such that the limit of κ_s as $s \to a$ is an incomplete geodesic, or there exists a unique congruence generated by γ, and the members of this congruence have lengths (when lifted to horizontal curves in LM) less that 4ϵ.

Proof

We use the notation of lemma 2.1.1. Without loss of generality we can suppose that γ is the projection of a horizontal curve λ. Let U be an open neighbourhood of p in which $|R^i{}_{jkl}| < K$, say, and define balls centred on p by $W_\delta = \{q \mid d(p,q) < \delta\}$. Then choose δ so small that $W_\delta \subset U$ and $\delta < \frac{1}{2}K^{-1/2}$. Then we take ϵ to be $\delta/20$.

We take the timelike curve γ to be specified by a parameter proportional to the generalised affine parameter, but scaled so that the length of the tangent vector, in a frame corresponding to a horizontal lift ending at p, is $\delta/20$, so that the range of the parameter is a subset of $(0,1)$. Then a congruence will be generated by γ, and will be unique, at least up to the first point at which either the congruence tends to an incomplete causal geodesic, or there is a violation of one of the inequalities

$$k = \|Y\| < \delta/5 \qquad\qquad (i)$$

or

$$\|L\| < 2. \qquad\qquad (ii)$$

Up to this point we can verify immediately that the conditions of lemma 2.2.4 are applicable, and the conclusion of the lemma gives the result that $\|L\| < b$, where $b < 2$. Thus, if there is to be a violation of one of the above inequalities it will be (i) rather than (ii). Moreover, we can calculate k by means of the relations

$$\frac{d\|Y\|}{dx} = -Y.\frac{dY}{dx} = -Y.\nabla_Y X = -Y.X|_{y=1}$$

since it is easily verified from the Riemann symmetries applied to Jacobi's equation that $Y.X$ is linear in y. Using the fact that at $y = 1$ the vector X has length $\delta/20$ in the F-frame, together with (i), allows us to conclude that in fact (i) cannot be violated, and the result follows. $\square$

Now we turn to properties of a topological nature.

6.2 Global hyperbolicity

6.2.1 Basic concepts

In a two dimensional vector space, if a quadratic form g has signature $(-,+)$, then the curves

$$g_{\alpha\beta}X^\alpha X^\beta = \text{const.}$$

are hyperbolae, while for signature $(+,+)$ they are ellipses; more generally, in n dimensions the surfaces given by this equation are hyperboloids when the signature is $(-,+,+,\cdots,+)$, as is the case for space-time. So this signature is referred to as hyperbolic. Correspondingly, a second-order partial differential equation is referred to as hyperbolic if its leading term has the form $g^{\alpha\beta}u_{,\alpha}u_{,\beta}$ with g of hyperbolic signature.

The main characteristic of hyperbolic equations is the local existence (i.e. existence in a sufficiently small region) of solutions to the Cauchy problem, the literature on which we have already referred to in section 4.4. In general terms, the result is that if one restricts attention to a sufficiently small region U, and if in this region there is prescribed Cauchy data (the value of u and of a derivative of u in a timelike direction) on a spacelike hypersurface S, the there exists a uniquely determined solution in a region called the domain of dependence of S in U, $D(S,U)$. The definition is that $D(S,U) = D^+(S,U) \cup D^-(S,U)$, where $D^+(S,U)$ (resp $D^-(S,U)$) is the set of all points p such that every past-directed (resp. future-directed) inextendible timelike curve from p intersects S.

A globally hyperbolic space-time M is one where the domain D in which the solution is determined by Cauchy data can be extended to the entire space-time; in the sense that there is an achronal surface S (a surface such that no two points on the surface can be connected by a timelike curve) for which $M = D(S,M)$.

6.2.2 Spaces of curves

While this definition is one of the more intuitively appealing ones, it differs from the original historical definition, which is in fact the one that will be used most often here, involving spaces of curves. To present this, we first review some basic ideas about sets of maps.

Suppose X and Y are topological spaces, and let $F(X,Y)$ be the set of continuous maps from X to Y. There are two different sorts of topology that can be put on this set. For the first, we note that given $f \in F(X,Y)$ there is determined a point $f(x)$ in Y for each x and so f can be regarded as a member of the infinite product $\prod_{x \in X} Y_x$, where the spaces Y_x are an infinite set of copies of Y, one for each element of X. The natural topology (called the Tychonoff topolgy) on an infinite product is constructed by taking as the open sets all unions of sets of the form $\prod_{x \in X} U_x$, where for each x, $U_x \subset Y_x$ and, for all but a finite number of values of x, $U_x = Y_x$. In terms of convergence, this means that $f_n \to f$ iff, for each x, $f_n(x) \to f(x)$.

The trouble with the Tychonoff topology is that the convergence expressed in the last equation may be non-uniform (e.g. consider the functions

$$f_n(x) = \begin{cases} nx & (0 \le x \le 1/n) \\ 2 - nx & (1/n < x \le 2/n) \\ 0 & (2/n < x \le 1\ ) \end{cases}$$

from $[0,1]$ to itself). To improve on this, one can use the compact-open topology, where the open sets are unions of finite intersections of sets of the form $H(T,U)$, for T compact in X and U open in Y, defined by

$$H(T,U) = \{f \mid f(T) \subset U\}.$$

If Y is a metric space, with metric d, then there is yet another topology available, the C^0 topology, where we require for convergence of f_n to f that $d(f(x), f_n(x)) \to 0$ uniformly in x. Fortunately, if X is compact this coincides with the compact-open topology.

We are interested in the case where $X = [a,b]$, $Y = M$ and $F(X,Y)$ is a space of continuous curves in M. Let $C(p,q)$ denote the subspace of F consisting of all causal curves f with $f(a) = p$ and $f(b) = q$. This subspace will then derive a topology from the

Tychonoff topology on F, and also a topology from the compact-open topology on F. But if we are dealing with causal curves, these topologies coincide (Beem and Ehrlich, 1981) (i.e. all convergence is uniform), so that there is no ambiguity as to the topology on all causal curves parametrised by a fixed closed interval.

Since the parametrisation is arbitrary (and cannot be fixed as proper time, because that would prevent us using a fixed interval for the parameter range) we must now identify curves that differ by a reparametrisation. In other words, we put on the space $C(p,q)$ an equivalence relation $\sim$ corresponding to reparametrisation, and move to the quotient space $C(p,q)$ under this relation. This has a natural topology induced by the map α from $C(a,b)$, namely the topology whose open sets U are precisely those for which $\alpha^{-1}(U)$ is open.

We note in passing that, now that we are dealing not with curves but with equivalence classes of curves under reparametrisation, we could regard such a class as simply a set of points in M and look for a topology on the collection of all such sets. Given a collection $\{A_n\}$ of subsets of M we define the Hausdorff lower limit to be the set

$$\liminf\{A_n\} = \{p \in M \mid \text{each neighbourhood of } p \text{ intersects all but a finite number of the } A_n\}$$

This gives rise to the idea of a limit curve: a curve f is said to be a limit curve of a sequence $\{f_n\}$ if $f([a,b]) \subset \liminf\{f_m([a,b])\}$ for some subsequence $\{f_m\}$ of the f_n. Beem and Ehrlich (1981) give further details of this idea, and they show that if (M,g) is strongly causal, then convergence of causal curves in the C^0 topolgy can be expressed in terms of limit curves. It turns out that, in the strongly causal case, if $f_n(a) \to p$ and $f_n(b) \to q$, then a causal curve f with $f(a) = p$ and $f(b) = q$ is a limit curve of the f_n iff there is a subsequence of the f_n that tends to f in the C^0 topology.

Using the topology just defined on $C(p,q)$, we finally have the result (due to Geroch – see (Hawking and Ellis, 1973)) that M is globally hyperbolic iff M is strongly causal and $C(p,q)$ is compact for all p, q with $p \in I^-(q)$. This is the original and useful form of global hyperbolicity.

6.2.3 Further results

The reader is referred to (Hawking and Ellis, 1973) for all details
in what follows. Perhaps the most important result, again due to
Geroch, shows that there is a homeomorphism ϕ from M to $S \times \mathbb{R}$,
where S is a three-manifold, such that each surface $\phi^{-1}(S \times \{a\})$
(for $a \in \mathbb{R}$) is a Cauchy surface. The function $t = \pi \circ \phi$ where
π is projection on the $\mathbb{R}$-component, defines in this case a time-
coordinate on the space-time, which runs from $-\infty$ to $+\infty$ on
every inextendible timelike curve. Of course, by rescaling t we
could equally arrange for the time-coordinate to run over the range
$(-1, 1)$, say, with time increasing in the future direction.

It is clear that the existence of a time coordinate with these
properties is equivalent to global hyperbolicity. However, one can
have space-times (such as the region $x > 0$ in Minkowski space, for
example) in which there exists a coordinate t which increases on
every future-directed inextendible timelike curve, but which does
not run over its maximum range on all timelike curves. Such a
space-time is not globally hyperbolic, though it is stably causal:
it is free from closed timelike curves with the given metric and for
all sufficiently close metrics.

6.2.4 Global hyperbolicity and singularities

Since global hyperbolicity is closely linked with questions of causality, it is natural to use a form of causal boundary in connection
with globally hyperbolic spaces. So we shall use the boundaries
$\dot{M}_i^{\pm}$ introduced in 3.3., recalling that each point p of $\dot{M}_i^{\pm}$ is an
equivalence class of incomplete inextendible past- (resp. future-)
directed causal curves γ having the same set for $I^{\pm}(\gamma)$. We denote
this set by $|p|$. The following result is then easy to demonstrate.

Theorem 6.2.1 *If* (M, g) *is globally hyperbolic and* $p \in \dot{M}_i^{+}$,
$q \in M$, *then* $|p| \not\subset I^{+}(q)$.

Proof

Suppose that (M, g) is globally hyperbolic. Take a $\gamma \in p$, $\gamma :$
$[0, a) \to M$, and suppose (to get a contradiction) that $|p| \subset I^{+}(q)$
for some q; in which case $\gamma(s) \in I^{+}(q)$ for each s so that there is

a past directed timelike curve $\gamma_s : [0, a] \to M$ with $\gamma_s(0) = \gamma(s)$, $\gamma_s(a) = q$. Define a curve κ_s for each s by

$$\kappa_s(t) = \begin{cases} \gamma\left(\frac{2st}{a}\right) & (0 \leq t < a/2) \\ \gamma_s(2t - a) & (a/2 \leq t \leq a). \end{cases}$$

These constitute an infinite set of curves in the compact set of curves $C(q, \gamma(0))$, and so have an accumulation curve κ. For $t < a/2$,

$$\kappa(t) = \lim_{s \to a} \tilde{\kappa}_s(t)$$

where $\tilde{\kappa}_s$ is a reparametrisation of κ_s. But since the curves κ_s agree on their initial segments, coinciding with γ there, it follows that initially κ is a reparametrisation of γ, contradicting the in-extendibility of γ. $\square$

Obviously this theorem has a dual in which past and future are interchanged.

If it had been the case that

$$|p| \subset I^+(q),$$

then the past-incomplete curve γ would have been entirely to the future of the regular point q. An observer travelling along one of the curves κ_s (in the future direction) would be in the past of the singularity at q but in the future of the singularity at $\gamma(0)$; and in a sense the curves κ_s, as s increases, get arbitrarily close to the singularity. Thus a singularity of this kind is fully open to inspection by observers.

6.2.5 The strong cosmic censorship hypothesis

There are grounds for believing that at least some "sufficiently strong" singularities (see (Newman, 1984) and chapter 8) do not have the "inspectability" property just described. These singularities are "censored" in that the geometry of space-time is such that they cannot be inspected. We formulate this as two definitions:

A singularity $p \in M^+$ (resp. M^-) is *strongly censored* if, for every $q \in M$, $|p| \not\subset I^+(q)$ (resp $I^-(q)$).

A space-time (M, g) *has strongly censored singularities* if all singularities of M^+ and M^- are strongly censored.

The hypothesis naming this section actually goes further than simply asserting that plausible sorts of space-times have strongly censored singularites; following the dictum of Polya that the more general a result is, the easier it is to prove, the hypothesis is that every generic, inextendible space-time containing physically reasonable matter is globally hyperbolic. By the theorem just proved, this implies that all the singularities are strongly censored.

The word 'strong' in this is to distinguish the hypothesis from a weaker version, which asserts that, under suitable reasonableness conditions, in an asymptotically flat space no singularities are visible from infinity.

There is another way of looking at the situation that obtains under the strong cosmic censorship hypothesis, by using the time coordinate t that exists in a globally hyperbolic space-time. Since this runs over its full range on al timelike curves, we see that all the singularities in M^+ are reached as $t \to -\infty$ along causal curves to the past, while all those in M^- are reached as $t \to +\infty$. We have a complete division of those singularities that are associated with incomplete causal curves into a past "big bang" and a future "big crunch", with nothing in between.

6.2.6 The significance of global hyperbolicity

At present the evidence is tending against the strong cosmic censorship hypothesis, in that increasingly more solutions are becoming known which are not globally hyperbolic and which, while not being either generic or containing completely physical matter, are still not too unreasonable from a physical point of view (Newman, 1986). But even if it turns out that space-time is not globally hyperbolic, it remains true that the part of space-time that is predictable, namely the part where the metric depends uniquely on data on a Cauchy surface, is by definition globally hyperbolic. Thus global hyperbolicity becomes a useful, indeed a vital, concept for understanding how a space-time evolves from initial data. Global hyperbolicity may break down because of the formation of a singularity, because of causality violation or because of the formation of a "timelike infinity" (or any combination of these). One approach is then to demand that the development of the data is maximal and examine any singularity that may form on the future boundary of the globally hyperbolic region (as we shall do below in section 7.2.1). These are the only singularities that are strictly determined by the initial data. But it turns out that there is a wider class of singularities that have the same essential properties as singularities in globally hyperbolic space-time, defined in terms of a weaker concept than global hyperbolicity, not depending on the cosmic censorship hypothesis.

6.3 Past hyperbolicity

6.3.1 Basic ideas

A globally hyperbolic region has a particularly simple structure (a direct product of space and time) which makes it easy to analyse the nature of singularities occurring on its boundary. But to apply this simplicity to singularities it is not necessary to have the whole of space-time globally hyperbolic, or even for there to exist a partial Cauchy surface (a spacelike surface without an edge). The minimal requirement is that the region immediately to the past of the singularity (if it is a singularity accessible by a future-directed curve) should be globally hyperbolic. The problem is to define a set S which can be regarded as "immediately to the past" of p. If p is the endpoint of a future directed curve $\gamma : [0, 1) \to M$ then a natural candidate is $T(t) := I^-(\gamma) \cap I^+(\gamma(t))$ for large enough t. But the sort of difficulty that might then arise is illustrated by the Taub-NUT-like example illustrated below, where $T(t_0)$ is the set shaded in the figure.

It will turn out that we need a way of restricting attention from the whole of this set to a smaller set which in the example is in a sense the set that would be obtained, were it not for the "topological" identifications that have taken place. We formalise this in the next definition.

6.3.2 The unfolding of a quasi-regular singularity

Let γ be a future timelike curve with a fixed horizontal lift $\bar{\gamma}$. We suppose the initial point of γ is chosen so that the generalised affine parameter length of γ is less that $\epsilon/20$, where ϵ is as in Proposition 6.1.3. Define

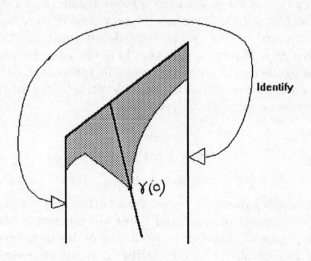

Fig. 2. The set $I^+(\gamma(0)) \cap I^-(\gamma)$.

$\exp_s \; := \exp |T_{\gamma(s)}M,$

$U(s) := \{X \in T_{\gamma(s)}(M) \,|\, X \text{ is future-pointing timelike},$

$\qquad\qquad\qquad \exp(X) \text{ is defined}$

$\qquad\qquad\qquad \operatorname{rank}((d \exp_s)(tX)) = 4, \forall t, 0 \le t \le 1\}.$

Then $U(s)$ will be a space-time if we give it the metric $(\exp_s)^*g$ pulled back from M.

Definition. A singularity p is said to be *past hyperbolic* if it is the endpoint of a future directed causal curve γ such that, for all large enough s,

1. there exists a curve $\gamma_s' : [s, 1) \to (U(s))$ with $\exp(\gamma_s'(t)) = \gamma(t)$;
2. there exists a globally hyperbolic subspace K of $U(s)$ with

$$U(s) \supset K \supset \overline{U_0(s)}$$

where the closure is with respect to the relative topology on $U(s)$ and we define

$$U_0(s) = I^-(\gamma_s'; U(s)).$$

In this case the set $U_0(s)$ will be refered to as the *unfolding* of the space-time near the singularity. Note that, since a past subset of a globally hyperbolic space is again globally hyperbolic, the unfolding is globally hyperbolic. The main role of the unfolding is to provide a space-time in which we can control the length of geodesics. This is done by identifying the frame $\bar{\gamma}(s)$ (for a given lift of γ) with a frame at $\gamma_s'(s)$ in $U(s)$ (using the identification of a vector space with its tangent space at the origin) and then propagating this frame along geodesics from the origin to obtain a frame-field e in $U(s)$. Any geodesics κ in $U(s)$ can then be measured by taking the generalised affine parameter length with respect to lift $\bar{\kappa}$ with $\bar{\kappa}(0) = e(\kappa(0))$. We denote this length measure by ℓ.

The following proposition, together with the implications that are obtained in conjunction with the results of chapter 2, are central to the next two sections.

Proposition 6.3.1 *For all timelike geodesics κ in $U_0(s)$, $\ell(\kappa) < \epsilon/2$; also geodesics in $U_0(s)$ have no conjugate points.*

Proof

Consider first timelike geodesics $\lambda : [0, 1] \to M$ ending on γ_s' (i.e. $\lambda(1) = \gamma_s'(t)$, say). Suppose we had $\ell(\lambda) > \epsilon/2$. Let ρ_r be the geodesic consisting of the straight line from the origin to $\lambda(r)$ (recall that $U_0(s)$ is a vector space, and by construction the radial curves are geodesic). Applying the results of chapter 2 to the family $\{\rho_r\}$ shows that $\ell(\rho_1) > \epsilon/4$. But ρ_1 is the member κ_t in the family generated by γ' according to Proposition 6.1.1, and so it must have a length $< \epsilon/5$ – a contradiction.

The argument for general geodesics $\sigma : [0, 1] \to U(s)$ proceeds in the same way: we first join $\sigma(1)$ to γ' by a geodesic σ_1 and then consider the family produced by joining the origin to the concatenation of σ_1 and σ. The absence of conjugate points follows immediately from the estimate on lengths and the results of Chapter 2 on Jacobi fields. $\qquad\square$

We note the following result, which is essential for the consistency of the terminology:

Proposition 6.3.2 *In a globally hyperbolic space-time, all quasi-regular singularities are past-hyperbolic.*

The Proof is straight forward from the definitions.

6.3.3 Primordial singularities

To understand the significance of past hyperbolicity let us suppose that the space is such that no boundary point is past hyperbolic. We can then show that there is a chain of boundary points stretching back into the past without a first member. To analyse this idea, we use the version of the causal boundary described in 3.3 , in which M^* denotes the collection of sets of the form $I^-(\gamma)$ for a causal curve γ, whether γ is incomplete or not (IPs) and M_i^* is the subset arising from restricting M^* to *incomplete* curves.

We recall that we called a set S in M^* a *causal chain* if the partial order $\subset$ when restricted to S was a (total) order. Define $p < q$ for $p \subset q$ and $p \neq q$, and for any ordered set T, and subset U of T write $(U)_T$ for

$$\{x \in T \mid \exists p, q \in U, p < x < q\}.$$

Call S *internally maximal* if there exists no causal chain $S' \subset M^*$ of which S is a proper subset having $(S' \backslash S) \subset (S)_{S'}$. Finally call a chain S *dense* if, for any $p, q \in S$ with $p < q$ there exists an $r \in S$ with $p < r < q$.

We can then define a *causal path* to be a set S in M_i^* such that:

1. S is a causal chain,
2. S is dense,
3. S is internally maximal.

From now on we assume that M is strongly causal, so that points in M are distinguished by their pasts. Note that we do not require that a causal path should be homeomorphic (with respect to the $\subset$ -topology) to a curve: this is because in practical applications we cannot avoid it being punctuated by points in $\bar{M}_i \backslash M_i^*$. But for singular causal paths we can include these easily: given any causal chain S in $M_i^* \backslash M$ we can form the causal completion

$$\bar{S} := \{p \in M^* \mid (\exists T \subset S)(p = \sum T)\} \subset \bar{M}_i$$

(noting that every subset T of S is itself a chain). $S \subset \bar{S}$ because for any s in S we have $s = \sum\{s\}$. As we noted in 3.3, if for some $q \in S, T \subset S$ we have $q \supset r$ for all $r \in T$, then $q \supset (\sum T) \supset r$ (for all $r \in T$) and so $\bar{S}$ is a causal chain (with respect to the order defined by inclusion). Then it turns out that a causal path can be realised by a curve mapping into the completion, in the sense of the following Proposition.

Proposition 6.3.3

If S is a causal path in $M_i^*\backslash M$, then there exists an order-isomorphism $\kappa : I \to S' \subset \bar{M}_i$ from an (open, half-open or closed) interval $I \subset \mathbb{R}$ (with the natural order) to a causal chain S' with $S \subset S' \subset \bar{S}$. Moreover, κ is maximal in the sense that there is no $\subset$-order-preserving $\kappa' : I' \to M_i^*$ with a $J \subset (I)_I$ such that $\kappa'(I'\backslash J) = \kappa(I)$ and $\kappa(I)$ a proper subset of $\kappa'(I)$.

Proof

Suppose S is a causal path. We first show that $\bar{S}$ contains a countable subset D dense in S – i.e. between any two distinct members of S there is a member of D. Take a countable dense subset $\{x_n \mid n \in \mathbb{Z}\}$ of M (such a set exists because M is paracompact). Let $\mathbb{Z}'$ be the subset of $\mathbb{Z}$ such that for each $n \in \mathbb{Z}'$ the set

$$S_n := \{s \in S \mid x_n \notin s\}$$

is non-empty and define $s_n := \text{l.u.b.} S_n$, for each $n \in \mathbb{Z}'$, $D := \{s_n \mid n \in \mathbb{Z}'\}$.

To show that D is dense, take any $p, q \in S$ with $p \neq q$. Since S is a chain we can suppose without loss of generality that $p < q$. Using the denseness of S, pick $p', q' \in S$ with $p < p' < q' < q$. Since the $\{x_n\}$ are dense we can choose an m such that $x_m \in q'\backslash p'$. Then $x_m \notin p'$ and so $m \in \mathbb{Z}'$ with $p' \in S_m$, i.e. $p' \leq s_m$. Moreover, it is impossible that $q' < s_m$, because in that case we should have an $s \in S_m$ with $s > q'$, implying that $x_m \notin q'$, a contradiction. So we have $p' \leq s_m \leq q'$ and denseness is proved.

Now let D' denote D without its greatest and least members (if any). D is dense in itself and so (Hocking and Young, 1961, p. 51) is order-isomorphic to $\mathbb{Q}' = \mathbb{Q} \cap (0, 1)$, the rationals in $(0, 1)$, by a map $\psi : \mathbb{Q}' \to D$. ψ can be extended to an order isomorphism

ψ' from $(0,1)$ to $\bar{S}$ as follows. Note that, since S is ordered by $\supset$, if the l.u.b. exists in S then it is unique. So for $x \in (0,1)$, if $t = \text{l.u.b.}_S\{\psi(\mathbf{Q} \cap (0,x))\}$ exists then define $\psi'(x) = t$; otherwise set $\psi'(x) = \sum\{\psi(\mathbf{Q} \cap (0,1))\}$. Set $S' = \psi'((0,1))$. It is straightforward to show that S' is correctly ordered by $\subset$ and contains S, apart possibly from the greatest and least members of S. But these can be restored by adding one or both of the endpoints of $(0,1)$.

Maximality is an immediate consequence of condition (3) in the definition of a causal path.   □

If a causal path lies entirely in the boundary of M^* then it can represent a singularity (if M is inextendible, so that all boundary points are singularities), and if it cannot be continued to the past then we have a singularity with no "beginning". Thus we make the following

Definition. A *primordial singularity* is a causal path S in $M_i^* \backslash M$, for an inextendible space-time M, such that S has at least two points and has no extension to the past as a causal path in M^*.

The main result of this section is now the following, which will be applied later to show that the members of a reasonably large class of space-times either have curvature singularities (i.e. singularities that are not quasi-regular) or primordial singularities.

Theorem 6.3.4

Suppose an inextendible strongly causal space-time (M,g) is such that

1. there exists a b-boundary singularity that is the endpoint of a future directed timelike curve, and every such singularity is quasi-regular;
2. no singularity in M_i^* is past hyperbolic.

Then (M,g) has a primordial singularity.

Proof

The idea is simply that condition (2) above implies the existence of boundary points (in M^*) arbitrarily close to, and in the past of, every singularity. If we can show that these boundary points (or at least a dense subset of them) are in fact in M_i^*, then we can build them up into a primordial singularity.

First we investigate the consequences of a failure of past hyperbolicity. It turns out that this produces singularities related by a causality condition stronger than that used previously, defined as follows. For TIPS X, Y we write

$$X \prec Y \qquad \text{if either (i)} \quad X \ll Y$$
$$\text{or (ii)} \quad X = I^-(\lambda), Y = I^-(\gamma)$$
$$\text{with some } s \text{ such that}$$
$$\lambda = \exp_s \circ \lambda' \text{ and } \lambda' \subset I^-(\gamma'; U_0(s))$$

A causal chain (or path) whose elements are $\prec$-related will be called *strictly causal*.

We can now prove

Lemma 6.3.5 *Let p be a quasi-regular singularity, the endpoint of a future timelike curve c, which is not past hyperbolic. Then for all sufficiently large s there exists a TIP $X \in M_i^*$ with $X \prec I^-(\gamma)$ and $\gamma(s) \in X$.*

Proof
Since p is not past hyperbolic, either (i) or (2) in the definition 6.3.2 must fail.
Failure of (1)
By considering the congruence generated by γ we see from 6.1.3 that γ' can only fail to exist if this congruence produces an incomplete geodesic κ not ending at p. In this case $I^-(\kappa) \subset I^-(\gamma(u))$ for large enough u, and hence $I^-(\kappa) \ll I^-(\gamma)$, i.e. $X = I^-(\kappa) \prec I^-(\gamma)$ and we are done.
Failure of (2)
We can suppose that (1) is satisfied, i.e. that γ' exists. So we have to prove for this case that $\sim (2) \Rightarrow \exists X$, or equivalently $\sim \exists X \Rightarrow (2)$. Thus let us suppose that there is no X of the form described in the lemma. Consider then some x with

$$x \in \text{boundary}\left(I^-(\gamma'; U(s)), T_{\gamma(s)}\right) \equiv A,$$

say. If $\exp x$ did not exist then for κ denoting the geodesic corresponding to x we would have $X = I^-(\kappa) \prec I^-(\gamma)$, contrary to hypothesis. So $\exp x$ exists for all such x and so since $U(s)$ is open it follows that $A \subset U(s)$. Hence it follows that the closure of $U_0(s)$ in $U(s)$ is contained in K, the domain of dependence of the past boundary (in $T_{\gamma(s)}$) of $U(s)$; for otherwise there would be generators of the horizon terminating in points of A not in $U(s)$. Thus K verifies (2)(a) as required. (2)(b) follows straight forwardly from the non-existence of X. □

Next we use this to build up causal paths.

Lemma 6.3.6

Under the conditions of the theorem, if S is a strictly causal chain in $M_i^* \backslash M$ such that there is no strictly causal chain S' in $M_i^* \backslash M$ properly containing S with $S' \backslash S \subset (S)_{S'}$, then S is dense.

Proof

Suppose given $q, r \in S$ with $r \prec q$, and suppose (in order to obtain a contradiction) that there exists no other point of S between q and r. Two cases arise, corresponding to the two cases in the definition of $\prec$.

Case a. Suppose there exists an x in M with $r \subset I^-(x) \subset q$. Let λ be an incomplete timelike curve for which $q = I^-(\lambda)$, and take s_0 such that $\lambda(s_0) \in I^+(x)$. Now apply lemma 1 to λ to give a TIP v with $v \prec q$ and $x \in v$. Then $r \ll v \prec q$ and so $S' := S \cup \{v\}$ is again a strictly causal chain and is larger than S. Thus S is dense.

Case b. Now suppose that $r = I^-(\lambda), q = I^-(\gamma)$ with

$$\lambda \subset \exp_s U(s)$$

and (2)a, b of the definition of $\prec$ satisfied. We work in $U(s)$ and define γ' as before, λ' being defined similarly so that $\exp_s \circ \lambda' = \lambda$. Let r', q' be the endpoints of λ' and γ' in $T_{\gamma(s)}$. Then we construct a point between r' and q' as follows. Choose s_1 so that $\gamma'(s_1) \notin I^-(\gamma'; U(s))$ (this being possible because otherwise q and r would be the same TIP). Choose a sequence $\{t_n\} \to 1$ and let κ_n be a timelike curve in $U(s)$ from $\lambda'(t_n)$ to a point in γ'. Suppose κ_n cuts $\dot{I}^+(\gamma'(s))$ in y_n. Then the $\{y_n\}$ have an accumulation point y. The geodesic κ corresponding to y generates a TIP $v = I^-(\kappa)$ satisfying the same conditions as in case (a). □

Lemma 6.3.7
Under the conditions of the theorem, if S is a causal path in M_i^*
with a least point q, then there exists a causal path S' extending S
to the past.

Proof
Let γ be a timelike curve ending at q. Applying lemma 1 in the
same way as in the proof of lemma 2, we can find a $y \in M$ ($y =$
$\gamma(s)$) such that there exists a point $p \in M_i^* \backslash M$ with $y \in p$ and
$p \prec q$; moreover, every point $p \in M^*$ that corresponds to a point
of $U_0(s)$ under exp and satisfies $y \in p$, $p \prec q$, is in M_i^*. Let T be a
maximal chain made up of such points p in $M_i^* \backslash M$ (which exists,
by Zorn's lemma and the fact that the union of an increasing
sequence of chains is a chain). From lemma 2 T is dense. Suppose
T were not internally maximal, i.e. that there was a $v \in M^*$ with
$p \prec v \prec q$ and $p, q \in T$. By continuity we must have $v = I^-(z)$
for $z \in M$. Construct a l.u.b for the set $E = \{r \mid v \prec r \prec q\}$ as
follows: z is clearly an internal point of the set

$$\prod E = \bigcap_{r \in E} r$$

and so the set

$$P = \left\{ I^-(w) \mid w \in \prod E \text{ and } z \in w \right\}$$

is non-empty. Choose a maximal causal chain P' in P and set
$k = \sum P$. k is then in $M_i^* \backslash M$ and so by lemma 1 we could find
a $u \in M_i^* \backslash M$ with $z \in u \prec k$. The addition of such a u would
contradict the maximality of T (as defined), showing that T is
internally maximal. Thus $S' = S \cup T$ is the required extension.
□

Proof of Theorem (contd.)
   Choose a timelike curve γ ending at a b-boundary singularity
(i.e. an incomplete inextendible timelike curve) and let $p = I^-(\gamma)$
be the corresponding point of M^*. Let V be the set of causal
chains in $M_i^* \backslash M$ (the boundary) that contain p and are internally
maximal (in the sense of (3) of the definition of a timelike path).
V is non-empty ($\{p\} \in V$) and partially ordered by inclusion; so
we can use Zorn's lemma to construct a maximal subset of V that

is linearly ordered by inclusion. The union of the elements of this subset constitutes a causal chain S containing p that is maximal over all internally maximal causal chains in $M_i^*\backslash M$.

We can show that S is a causal path if we verify that S is dense. The proof is identical to that in lemma 2, with the addition that the set S' there is internally maximal because if there were an extension $\bar{S}'$ in M^* satisfying the conditions on S' in the definition of internal maximality, then $\bar{S}'\backslash\{v\}$ would be such an extension of S, contradicting the internal maximality of S. So S is a causal path.

The last step is to show that S cannot be extended to the past in M^*. Suppose that $T \subset M_i^*$ were such an extension, T being a causal path with $T\backslash S$ to the past of S. Let $w = \sum(T\backslash S) \in M^*$. If w were not in T then $\{w\}\cup T$ would be a causal chain extending T internally, contradicting the internal maximality of T; so $w \in T$. We cannot have $w \in (T\backslash S)$, because than $S \cup \{w\}$ would be an enlargement of S, contradicting the maximality of S, so $w \in S$. But now lemma 3 gives an extension, again giving a contradiction.

$\square$

6.3.4 Dragging geodesics

We shall apply the construction of Proposition 6.1.3 frequently in the special case where the curve γ has the form of a geodesic κ followed by a timelike curve γ' concatenated to κ. In that case the part of the family generated by γ corresponding to κ consists simply of a family of segments of κ, which is then continued as a family of geodesics terminating on γ'. For shortness of terminology, where this context is understood we shall refer to this latter family alone (omitting the family of segments of κ) as the family generated by γ'.

The construction of generating a family of geodesics, in the region to the past of a past-hyperbolic quasi-regular singularity, allows us to "drag" geodesics along timelike curves. Suppose given a geodesic $\kappa : [0, 1] \to T(s)$ and a timelike curve $\lambda : [0, 1] \to T(s)$ with $\lambda(0) = \kappa(1)$. Then we can form a family $\mu(s, t)$ generated by κ, λ with $\mu(s, 0) = \kappa(s)$, $\mu(1, t) = \lambda(t)$. We refer to $\mu(\,\cdot\, , 1)$ as the geodesic obtained by dragging κ along λ.

A second application is to the timelike curve γ defining a quasi-regular singularity. We can construct the family generated by the part of γ to the future of $q = \gamma(s)$ for s sufficiently near 1: this gives a family κ_t of geodesics with $\kappa_t(0) = q$, $\kappa_t(1) = \gamma(t)(s \leq t \leq 1)$. The proof of the existence of the generated family shows that the initial tangent vector $\dot{\kappa}_t$ tends to a timelike vector $\dot{\kappa}_1$, from which it is easy to show from past hyperbolicity and quasiregularity that the geodesic κ_1 with this as initial tangent vector terminates at p.

For convenience of notation, let us now write κ for κ_1. The limiting geodesic of the family generated by γ will be a timelike geodesic ending at the same singularity. So from now on we can assume that the generating curve γ for a past-hyperbolic quasi-regular singularity is a timelike geodesic.

6.4 Past simplicity

Let p be a singularity at the end of a timelike curve γ, with $p = \lim \gamma(s)(s \to 1)$. The condition that p is past hyperbolic does not guarantee that its past resembles the past of a point: we have already seen in 6.3.1 that identifications can occur near the singularity, and we introduced the idea of the unfolding of a singularity to take care of this. The basic example to bear in mind in what follows is the so-called "two dimensional Taub-NUT space" defined by taking the part of two dimensional Minkowski space

$$t < 0, \alpha t \leq x \leq -\alpha t$$

(α constant, $0 < \alpha < 1$) with the boundaries identified. Here any part of the space-time close to the singularity at the origin is a cylinder, and the exponential map defined on any unfolding $U(s)$ is not 1-1.

Definition. We shall say that p is *past simple* if the conditions for past hyperbolicity (6.3.1) are met, with the addition that $\exp_s$ is 1-1 on K.

It seems likely that this is equivalent to the simple connectedness of slices of $I^-(\gamma)$, but no general proof of this is known.

6.4.1 Limiting Riemann sets

While these definitions gives a helpful picture for understanding what is going on geometrically, they are of limited use in practice because we have no a priori ground for supposing that a singularity should be past-simple. This next proposition establishes this for quasi-regular singularities that are generic, in the sense that the Riemann tensor is not algebraically specialised at p.

To define this last idea, take a frame $\mathbf{E}$ at p and let $\mathcal{R} = \mathcal{R}(\mathbf{E})$ be the limit set of the Riemann tensor components near $\mathbf{E}$; i.e.

$$\mathcal{R} = \{\{T^i{}_{jk\ell} \mid i, j, k, \ell = 0, \ldots, 3\} \mid (\exists\, \mathbf{F}_m)(\mathbf{F}_m \to \mathbf{E}$$
$$\&\, R^i{}_{jk\ell}(\mathbf{F}_m) \to T^i{}_{jk\ell})\}$$

While this is adequate for a continuous Riemann tensor, in the case where the metric is only C^{2-} this definition must be altered because the Riemann tensor is then only defined almost everywhere, and different choices of the Riemann tensor, differing on sets of measure zero, would give different limit sets.

Let $\mathbf{E} \in \overline{LM}$ be a point in the closure of the frame bundle. We define the essential limiting Riemann set $\mathcal{R}^*(\mathbf{E})$ as follows:

$T \in \mathcal{R}^*(\mathbf{E}) \iff$ For every neighbourhood V of $\mathbf{E}$ in $\overline{LM}$ and every $\epsilon > 0$, there exists a non-negative continuous function v with compact support in LV such that

$$\left| T - \int v(\mathbf{E}')\mathbf{R}(\mathbf{E}')d\mu(\mathbf{E}') \right| < \epsilon$$

$$\int v(\mathbf{E}')d\mu(\mathbf{E}') = 1,$$

where μ is the measure on LM induced by the b-metric.

From this point of view a *curvature singularity* can be defined as a point p such that, for any $\mathbf{E} \in \pi^{-1}p$, $\mathcal{R}^*(\mathbf{E})$ is unbounded, while a *quasi-regular singularity* is one that is not a curvature singularity.

It is clear that $\mathcal{R}^*$ is a convex set, and if R is continuous then it is simply the convex hull of $\mathcal{R}$. Also, it can be checked that $\mathcal{R}^*$ is covariant under the Lorentz group, in the sense that

$$\mathcal{R}^*(\mathbf{E}L) = L\langle \mathcal{R}^*(\mathbf{E})\rangle := \{L\langle \mathbf{R}\rangle : \mathbf{R} \in \mathcal{R}^*(\mathbf{E})\}$$

pointed brackets denoting the action of L on the array of Riemann components $\mathbf{R}$ corresponding to changing basis by L. Consequently, if L belongs to $G(\mathbf{E})$, the singular holonomy group at $\mathbf{E}$, then from $\mathbf{E}L = \mathbf{E}$ we have that $L\langle \mathcal{R}^*(\mathbf{E})\rangle = \mathcal{R}^*(\mathbf{E})$; i.e. $\mathcal{R}^*$ is invariant under the singular holonomy group.

6.4.2 Specialisation

For brevity we shall call an array of Riemann components *specialised* if the Weyl tensor corresponding to the array has a repeated principal null direction, and that direction is an eigendirection of the Ricci tensor (and hence also an eigendirection of the energy momentum tensor defined from the Ricci tensor via Einstein's equations). The significance of this lies in the following:

Proposition 6.4.1

Suppose (L_n) is a sequence of Lorentz transformations with $\|L_n\| \to \infty$, and that T is an array of Riemann components such that the elements of $L_n\langle T\rangle$ and of $L_n{}^{-1}\langle T\rangle$ (the transforms of T by L_n and $L_n{}^{-1}$ acting in the usual way) are bounded. Then T is specialised.

Proof

It is simplest to convert the expressions to spinor form, under which L_n corresponds to an $SL(2, \mathbf{C})$ matrix A_n which we can factorise as $A_n = U_{1n}B_nU_{2n}$ with the Us unitary and

$$B_n = \begin{pmatrix} a_n & 0 \\ 0 & a_n{}^{-1} \end{pmatrix}$$

Moreover we can choose the factorisation so that $a_n \to +\infty$. The array of Weyl components is equivalent to the array $\Psi = (\Psi_0, \ldots, \Psi_4)$ of Weyl spinor components, which are bounded under the action of A_n. Since a unitary matrix does not alter the norm of vectors on which it acts, this means that $B_nU_{2n}\langle\Psi\rangle$ is bounded, and (considering $A_n{}^{-1}$) $B_{1n}{}^{-1}U_n^{\dagger}\langle\Psi\rangle$ is bounded. But

$$B_n\langle\Psi_0, \Psi_1, \Psi_2, \Psi_3, \Psi_4\rangle = (a^4\Psi_0, a^2\Psi_1, \Psi_2, a^{-2}\Psi_3, a^{-4}\Psi_4).$$

Hence $(U_{2n}\langle\Psi\rangle)_0 \to 0$ and $(U_{2n}\langle\Psi\rangle)_1 \to 0$. Consequently any spinor κ^A which is a limit point of the $U_{2n0}{}^A$ as $n \to \infty$ (and such a spinor must exist, by compactness) satisfies $\Psi_{ABCD}\kappa^A\kappa^B\kappa^C = 0$.

Similarly any spinor λ^A which is a limit point of the $U_{1n}{}^\dagger{}_0{}^A$ as $n \to \infty$ satisfies $\Psi_{ABCD}\lambda^A\lambda^B\lambda^C = 0$. Thus κ and λ are repeated principal null directions.

Unfortunately, there is no guarantee that they are distinct: an example where they are the same is provided by

$$A_n = \begin{pmatrix} 0 & -a_n^{-1} \\ a_n & 0 \end{pmatrix}$$

$$\Psi = (0, 0, \Psi_2\Psi_3\Psi_4)$$

The assertion about the Ricci tensor follows straight forwardly in the same way: if we change the spinor basis so that κ has $\kappa^A = \delta_0^A$ then we find that, for boundedness,

$$\Phi_{00'00'} = 0,$$

$$\Phi_{10'00'} = 0,$$

$$\Phi_{01'00'} = 0$$

(and components related by symmetry), whence

$$S_{\alpha\beta}n^\alpha n^\beta = 0, \quad S_{\alpha\beta}n^\alpha m^\beta = 0$$

(where S is the trace-free Ricci tensor and (l, m, n) is the usual null tetrad). □

6.4.3 Measures of specialisation

While the above concept of specialisation is what is really needed for our results, for the sake of simplicity we shall use a rather wider concept, corresponding only to the Algebraic specialisation of the Weyl tensor and a related condition on the Ricci tensor. What will be needed is a condition that these tensors are, so to speak, "uniformly algebraically general" in a neighbourhood of the singularity. If the Riemann tensor were continuous this would be guaranteed by the demanding that the tensors were general at the singularity; but we are interested also in the case where the Riemann tensor can be discontinuous. The easiest way to proceed in this case is to introduce a quantitative measure of algebraic specialisation, as is provided by the invariant $H = (\Psi_{ABCD}\Psi^{ABCD})^3/8 - 3(\Psi_{ABCD}\Psi^{CDEF}\Psi_{EFAB})^2/4$ (cf. Stephani et al. p. 59).

We find by direct calculation that

$$H = \Psi_0^3\Psi_4^3 - 12\Psi_0^2\Psi_1\Psi_3\Psi_4^2 - 18\Psi_0^2\Psi_2^2\Psi_4^2 + 54\Psi_0^2\Psi_2\Psi_3^2\Psi_4$$

$$- 27\Psi_0^2\Psi_3^4 + 54\Psi_0\Psi_1^2\Psi_2\Psi_4^2 - 6\Psi_0\Psi_1^2\Psi_3^2\Psi_4 - 180\Psi_0\Psi_1\Psi_2^2\Psi_3\Psi_4$$
$$+ 108\Psi_0\Psi_1\Psi_2\Psi_3^3 + 81\Psi_0\Psi_2^4\Psi_4 - 54\Psi_0\Psi_2^3\Psi_3^2 - 27\Psi_1^4\Psi_4^2$$
$$+ 108\Psi_1^3\Psi_2\Psi_3\Psi_4 - 64\Psi_1^3\Psi_3^3 - 54\Psi_1^2\Psi_2^3\Psi_4 + 36\Psi_1^2\Psi_2^2\Psi_3^2$$

We can then use this to estimate the relation between the size of the Lorentz transformed Weyl tensor and the size of the Lorentz transformation. Let us write

$$W(\alpha) = \min_{\|L\| = e^\alpha} \frac{\|L\langle\Psi\rangle\|}{\|L\|\|\Psi\|}$$

where $\|\Psi\| = (\Psi_{ABCD}\bar\Psi_{P'Q'R'S'}t^{AP'}t^{BQ'}t^{CR'}t^{DS'})^{1/2}$, with t being the SU(2)-invariant spinor having components comprising the identity matrix in the chosen frame. If $|\Psi|$ denotes the size of the largest component, then we clearly have that $|\Psi| \le \|\Psi\| \le 4|\Psi|$. Then if B_α denotes a boost with rapidity α, we have that

$$W(\alpha) = \min_{\Psi'} \frac{\|B_\alpha\langle\Psi'\rangle\|}{e^\alpha\|\Psi\|}$$
$$\ge \min_{\Psi'} \max\left(e^\beta\Psi_4', \Psi_3'\right) / \|\Psi\|$$

for any $\alpha > \beta$, where Ψ' runs over all Weyl spinors that are SU(2)-related to Ψ.

But, from the above, $|H| \le (595\Psi_4' + 289\Psi_3')(\|\Psi\|)^5$, so that

$$W \ge |H|\|\Psi\|^{-6}/(595e^{-\beta} + 289).$$

In particular, taking $\beta = 0$ we have

$$\|L\| < 884\|L\langle\Psi\rangle\|/(|H|\|\Psi\|^{-5}).$$

A similar argument can be carried out for the trace-free Ricci tensor. If we use the spinor form $\Phi_{AX'BY'}$ to form the invariants

$$t = \Phi_{AX'BY'}\Phi^{BY'AX'},$$
$$u = \Phi_{AX'BY'}\Phi^{BY'CZ'}\Phi_{CZ'}{}^{AX'},$$
$$v = \Phi_{AX'BY'}\Phi^{BY'CZ'}\Phi_{CZ'DW'}\Phi^{DW'AX'},$$

followed by the compound invariant

$$w = 9t^6 - 90t^4v - 68t^3u^2 + 288t^2v^2 + 144tu^2v - 24u^4 - 288v^3,$$

then it can be shown that this has the form

$$w = \Phi_{00'00'}r + |\Phi_{00'01'}|s$$

where r is a polynomial in the other components of Φ and s is a polynomial in the other components and in the phase-factor of $\Phi_{00'01'}$. A similar argument then shows that there exists an estimate of the form

$$\|L\| < M\|L\langle\Phi\rangle\|/(\|w\|\|\Psi\|^{-11})$$

for some constant M.

Presumably there exists a similar invariant involving both the Ricci and Weyl tensors expressing the alignment between their canonical tetrads but it does not seem to be known explicitly and so we shall confine attention to the invariants just mentioned. We therefore make the following definition:

A (regular or boundary) point p is said to be a *point of generic curvature* if the invariants H and w are bounded away from zero in a neighbourhood of any frame above p.

Theorem 6.4.2

If p is a past hyperbolic quasi-regular singularity and a point of generic curvature, then p is past-simple.

Proof

If a singularity p is not past simple, then for all choices of s and K in 6.3.2 the map $\exp_s$ is not 1-1 on K; i.e. there are distinct vectors X, Y in K with $\exp X = \exp Y$. Let A be the neighbourhood of p in which $|H| > H_0 > 0$ and $|w| > w_0 > 0$, say. Then we can choose K small enough to allow us to drag the geodesics defined by X, Y so that their endpoints lie in A. Since this can be arranged to happen arbitrarily close to p, we can, by choosing a sequence of points s_i with $\gamma(s_i) \to p$ with corresponding sets K_i, find sequences $X_i, Y_i \in T_{\gamma(s_i)}$ with $\exp(X_i) = \exp(Y_i) \in A$. The Riemann tensor can be controlled in magnitude throughout the construction. We can choose K_i small enough to give $\exp(K_i) \subset \exp(K_{i-1})$, with a natural identification of K_i as a subspace of K_{i-1}. In this sense we can regard all the K_i and $U(s_i)$ as subspaces of some fixed $U(s)$.

Let $\mathbf{E}_i$, $\mathbf{E}'_i$ be the results of propagating in M the tetrad on the lift of γ at s_i along X_i, Y_i respectively. Since p is quasiregular, $\mathcal{R}(\mathbf{E}_i)$ and $\mathcal{R}(\mathbf{E}'_i)$ are bounded. Thus, from the preceding results on H and w, if L_i is defined by $\mathbf{E}_i L_i = \mathbf{E}'_i$, then the sequence of the $\{L_i\}$ is bounded.

Consider the limit set of the L_i, defined as

$$\Lambda = \{L \in \mathbf{L} \mid (\exists \{t_n\})(L(t_n) \to L \ \& \ t_n \to \infty \text{ as } n \to \infty)\}.$$

where, for typographic convenience, we write $L(t_n)$ for L_{t_n}, and similarly for the $\mathbf{E}$s.

We know that $\Lambda \neq \emptyset$ because the set of all the L_i is bounded.

Clearly L_n is the holonomy transformation associated with the loop σ_n formed by the geodesics with tangents X_n, Y_n. From the boundedness of the L_n, the lengths of the loops σ_n (measured in a frame parallely propagated on γ) tend to zero, so $\Lambda \subset G(\mathbf{E})$, the singular holonomy group at $\mathbf{E}$.

First, we show that if $L_0 \in \mathbf{L}$ is some arbitrary rotation (i.e. $L_0 \mathbf{X} = \mathbf{X}$ for some timelike vector $\mathbf{X} \in \mathbb{R}^n$), then $L_0 \notin \Lambda$. For suppose we did have $L_0 \in \Lambda$, with a $\{t_n\}$ such that $L(t_n) \to L_0$. Take $\mathbf{X}$ to be past directed and set $X_n = \mathbf{E}(t_n)^* L(t_n).\mathbf{X}$, $X_n' = \mathbf{E}(t_n)^{*\prime}.\mathbf{X}$, where we now define tetrads $\mathbf{E}^*$ as for $\mathbf{E}$ but by parallel propagation in $U(s)$ rather than M. Taking coordinates in $U(s)$ we have that the components of X_n and X_n' approach each other as $n \to \infty$, and the coordinates of the points where they reside approach each other. So if we draw geodesics with initial vectors X_n and X_n', the points x_n and x_n' at which these geodesics intersect a cauchy surface C to their past tend to each other. But X_n and X_n' are mapped into each other, and hence so are x_n and y_n. But this then contradicts the fact that ϕ is a local homeomorphism.

Thus Λ contains an element L_0 that is not a rotation.

Consequently, the cyclic subgroup of the singular holonomy group generated by L_0 contains elements of arbitrarily large modulus. But, using the above results on algebraic specialisation, this then contradicts the facts that the Riemann tensor is bounded near p, (because p is quasi-regular) and that the set $\mathcal{R}^*(\mathbf{E})$ is invariant under this group. $\qquad \square$

6.5 Holes

6.5.1 Initial concepts

There is a sense in which all singularities are "holes" in space-time, in that curves that might be complete are punctured by them and rendered incomplete. But it is useful to distinguish between the case where the singularity arises as the inevitable result of some dynamic evolution, and the case where the singularity is gratuitous, in the sense that the immediately preceding evolution might not have resulted in a singularity; or, more precisely, that there is some other space-time containing the same section of evolution as occurs in the past of the singularity, but where the singularity does not occur. In this latter case, one feels that any singularity exhibited by the mathematical solution is an artifact, and that the solution with the singularity should be rejected in favour of the solution without the singularity. It is this sort of gratuitous singularity that we call a *hole*. The exact definition is as follows. A point p on the b-boundary $\partial_b M$ is a future (resp. past) hole if there exists

1. a globally hyperbolic subspace K of M
2. a future (resp. past) directed curve γ in K ending at p;
3. an isometry $\theta : K \rightarrow N$, where (N, g') is a space-time, such that $\theta \circ \gamma$ is extendible as a curve in $D(\theta(K))$.

The simplest example of a hole is simply a point removed from a space-time N, with $M = N \backslash \{p\}$. In the definition we can take K to be $D(S)$ for any spacelike surface S in M such that $p \in D^+(S, N)$. Such a point is, of course, not a singularity because the spacetime as a whole can be extended so as to remove p from the b-boundary. But one can modify the point-removal idea so as to produce a genuinely singular hole. From $\mathbb{R}^4$ with the standard Minkowski metric, delete the 2-plane $H = \{x = 0, t = 0\}$, and let M be the covering space of $\mathbb{R}^4 \backslash H$. Again, each b-boundary point corresponding to a point in H is a hole, but once the covering space has been taken there is no way of extending the space-time as a whole so as to remove the singularity.

A space-time with no holes is termed hole-free. Though both the examples of holes so far cited have violated global hyperbolicity, hole-freeness is a weaker concept than global hyperbolicity. To see the independence of the concepts, consider the space-time formed by removing the timelike 2-plane $L = \{x = 0, y = 0\}$ from $\mathbb{R}^4$ and taking the covering space of $\mathbb{R}^4 \backslash L$ so as to give a genuine singularity. This is not globally hyperbolic (the removed plane interrupts many timelike curves and so breaks the compactness of the set of timelike curves connecting p and q for p and q close to

L) but it is hole-free because there is no way of getting a Cauchy surface to the past of the singular set L.

The converse case, of a globally hyperbolic space-time that is not hole-free, can be illustrated by the region $|t| < 1$ of Minkowski space; but this is of course extendible (to the whole of $\mathbb{R}^4$) and so we are not dealing with a genuine singularity. If we require that all b-boundary points are singularities (i.e. that the space-time is maximal), then we can show that a globally hyperbolic space-time must be hole free. This is equivalent to the following:

Proposition 6.5.1

If a globally hyperbolic space-time has a hole p then it is extendible.

Proof

Let θ, N, K be as in the definition of a hole. Define $N' = D(\theta(K))$, $M' = (M \cup N')/\theta$ (i.e. the quotient of the disjoint union of M and N' by the equivalence relation

$$x \sim y \equiv x = y \text{ or } \theta(x) = y \text{ or } \theta(y) = x.)$$

M' has a natural topology induced by the projection π of $M \cup N'$ onto M', which has as a basis all sets $\pi(U)$ with U open in M or in N'. So M' is locally homeomorphic to $\mathbb{R}^4$ and will be a manifold provided it is Hausdorff. If we can prove it to be Hausdorff, it will then have a natural metric induced from M and N' and so be a space-time extending M.

We proceed by contradiction: suppose that distinct points r and q in M' are not Hausdorff separated, in that, for every pair V_1, V_2 of neighbourhoods of r, q, respectively, $V_1 \cap V_2$ is non-empty. Clearly this can only happen if $q \in (K)$, $r \in (\theta(K))$, (or vice versa). By choosing a sequence of pairs of neighbourhoods, with a point z_i in the intersection of each pair, we can find a sequence $\{z_i\}$ such that $z_i \to q$ and $z_i \to r$. By construction $z_i \in \pi(K)$, so we have pairs x_i, y_i with $x_i \in K$, $y_i = \theta(x_i)$, $z_i = \pi(x_i) = \pi(y_i)$, $\{x_i\} \to q$, $\{y_i\} \to r$. Since $q \in (K)$ we can find a past directed null geodesic segment γ from q in $H + (S)$ with initial tangent vector X. Let $X_i \in T(x_i)$ be chosen so that $\{X_i\} \to X$; take frames $\mathbf{E}_i$ at x_i converging to a frame $\mathbf{E}$ at q and set $Y_i = \theta_*(X_i)$, $F_i = \theta_*(E_i)$. With respect to the metric on the frame bundle, $\{E_i\}$ is a Cauchy sequence and hence so is $\{F_i\}$. Thus $\{F_i\}$ converges to some limit $\mathbf{F}$ at r. The components $\mathbf{X}_i = \{X_i{}^j \mid j = 0,\ldots,4\}$ of X_i with respect to $\mathbf{E}_i$ converge to $\mathbf{X}$, the components of X, and so $\{Y_i\}$ converges to $Y = X^j F_j$. Since $r \in D^+(\theta(S))$, the geodesic with tangent vector Y intersects $\theta(S)$, where S is a Cauchy surface for K, in $\theta(s)$, say. By continuity, for large enough i the geodesics with initial tangent vector Y_i also intersect $\theta(S)$, in $\theta(s_i)$ say.

Choose $h \in I^-(s)$, $k \in I^+(q)$. For large enough i, $x_i \in I^-(k)$, $s_i \in I^+(h)$ and so the geodesics γ_i from x_i to s_i with initial tangent vector $\mathbf{X}_i$ can be extended to timelike curves from k to h. By global hyperbolicity these curves lie in a compact set and so have a limit containing the limit geodesic of the $\{\gamma_i\}$, which is a segment of γ. This implies that γ must cut S, contradicting the hypothesis that it is a generator of $H^+(S)$. $\qquad\square$

Now, the results of 5.4 show that, subject to differentiability conditions, the set K for a quasi-regular singularity is always extensible. We have seen that, in the past hyperbolic case, K can be regarded as a subspace of M. Thus by taking this K in the definition of a hole we can see the following result (a sort of converse to the preceding theorem): *A quasi regular, past simple, past hyperbolic singularity is a hole.*

7

Extension theorems

In this chapter I shall describe various situations in which it is possible to extend through a boundary point; in these cases the boundary point is not a singularity. As has been explained, we are proceding by elimination, so that the remaining cases must either be regarded as genuine singularities, or be amenable to extension by more powerful means than used here. There is no absolute criterion for what sorts of extension are "legitimate", and hence no absolute criterion for what is, and what is not, a singularity.

7.1 Spherical symmetry

In this situation (which is of considerable interest because of the ease of obtaining exact solutions) it is possible to prove the existence of extensions under weaker assumptions than is normally the case. The results are thus not only of interest in their own right, but may be an indication of the "best possible" results that might be obtainable in the general case.

7.1.1 Definition of the problem

We shall be dealing with a space-time in which the rotation group $SO(3,\mathbb{R})$ acts transitively on spacelike 2-surfaces. So through every point p of the space-time there exists in a neighbourhood of p a totally geodesic timelike 2-surface S orthogonal to the orbits of the group; the surfaces maximal with respect to these properties through a given group orbit are equivalent.

If e, f are two mutually orthogonal vectors at p tangent to the group orbit (and hence orthogonal to S) then the tensor of projection into S is defined by

$$h_{\alpha\beta} := g_{\alpha\beta} - e_\alpha e_\beta - f_\alpha f_\beta$$

We can distinguish various curvature quantities. First,

$$K = h^{\mu\nu} h^{\rho\sigma} R_{\mu\rho\nu\sigma} / 4$$

which is just the Gaussian curvature of **S**. Then

$$L_{\alpha\beta} = h_\alpha{}^\mu h_\beta{}^\nu e^\rho e^\sigma R_{\mu\rho\nu\sigma},$$

can be regarded as a tensor on **S**, where e is any unit vector orthogonal to **S** ($L_{\alpha\beta}$ being independent of the choice of e because the spherical symmetry renders all such vectors equivalent). Contraction gives the invariant

$$L = h^{\alpha\beta} L_{\alpha\beta}.$$

Now suppose that **S** contains a null geodesic γ that is incomplete in **S** (and hence in M) and inextendible in M (and hence in **S**). For the sake of definiteness we choose the time-orientation (at least locally) so that γ is future orientated. Let κ be a future directed null geodesic with affine parameter v cutting γ and let γ_v be the null geodesic transverse to κ through $\kappa(v)$, with the same direction of parametrisation as γ.

We now restrict attention to the situation where the singularity sufficiently near the endpoint of γ is a simple achronal surface in the following sense:

1. for appropriate choice of parametrisation of κ and for some sufficiently small $\epsilon > 0$ the geodesics $\gamma_v (0 < v < \epsilon)$ are incomplete (the original γ being γ_0).

2. If we define

$$V = \bigcup_{0 \le v < \epsilon} |\gamma_v|$$

(where $|\gamma|$ denotes the image of γ) then we can assign coordinates u, v to V such that each γ_v corresponds to $v = $ const., and $u = $ const. corresponds to the null geodesics κ_u transverse to γ through $\gamma(u)$ (with $\kappa = \kappa_0$). The coordinates are chosen so that u is the affine parameter on γ_0 and v is the affine parameter on κ, with u, v increasing to the future.

3. There exists a non-increasing function ϕ such that the range of coordinates on V is the domain U given by

$$0 < u < u_0$$
$$0 \le v < \phi(u).$$

(If we take the geodesic κ to be past directed then the same conditions define an *aspatial* singularity for which the results go through equally well).

Note that this condition restricts the "wiggliness" of the singularity.

We assume that the singularity on γ is weak in the sense that for v in a small enough interval about 0 the above curvature quantities are integrable with uniform bounds on finite affine-parameter intervals on the null geodesics γ_v and κ_u, i.e. that

$$\int_a^b |K(\kappa_u(\lambda))|d\lambda < M_K, \qquad \int_a^b |K(\gamma_v(\lambda))|d\lambda < N_K \qquad (1)$$

and similarly for L, and

$$\int_a^b |L_{\alpha\beta}(\gamma_v(\lambda))\dot{\gamma}^\alpha\dot{\gamma}^\beta|d\lambda < N_\gamma, \qquad \int_a^b |L_{\alpha\beta}(\kappa_u(\lambda))\dot{\kappa}^\alpha\dot{\kappa}^\beta|d\lambda < N_\kappa \qquad (2)$$

for all intervals $[a, b]$ of the affine parameter for which the geodesic in question is in V.

7.1.2 Extension in null coordinates

We shall now show that, under the assumptions just described, the metric has a C^{1-} extension; more precisely, we shall show that the components of the metric in (u, v)-coordinates can be extended through the line $v = \phi(u)$ in the coordinate plane.

By assumption, a neighbourhood of γ is covered by the geodesics κ_u and γ_v and so has been given null coordinates u and v. With respect to these coordinates the metric takes the form

$$2H\,du\,dv + \rho^2\left(d\theta^2 + sin^2\theta d\phi^2\right)$$

(with H, ρ functions of u and v) and we have

$$K = H_{,u}H_{,v}/H^3 - H_{,uv}/H^2 \qquad (3)$$
$$L = -H^{-1}\rho^{-1}\rho_{,uv}. \qquad (4)$$

Now it can be shown that, because K is the Gaussian curvature and has bounded integral on the geodesics γ_v, the ratio of the tangent vector to γ_v with respect to the affine parameter to the tangent vector with respect to the coordinate u is bounded, and similarly for v. (These parameters are equal on the geodesic γ_0.) Thus the integrals of K and L in (1) and (2) are bounded with respect to u and v. We use this to enable us to estimate H and ρ by regarding (3) and (4) as partial differential equations determining these quantities.

First, introduce $p = \ln H$, allowing us to rewrite (3) as

$$p_{,uv} = -Ke^p. \tag{5}$$

We shall describe only the achronal case: the aspatial case differs in the choice of signs.

The condition that u is an affine parameter on γ_0 implies (by calculating the Christoffel symbols) that $H_u = 0$ on γ_0, i.e. that p is constant here; similarly p is also constant on κ and by continuity the two constants are equal and can be made equal to unity by a suitable choice of one of the affine parameters. Then (5) can be rewritten as the integral equation

$$p(u,v) = -\int_0^u du' \int_0^v dv' K(u',v)e^{p(u',v')} + 1.$$

From the boundedness of the integral of K with respect to u we obtain

$$|p(u,v) - 1| < vuN_K \exp\left(\max_{\substack{0<u'<u \\ 0<v'<v}} |p(u',v')|\right)$$

Hence we can ensure that $p < 2$ provided that we choose ϵ originally less than $1/(4e^2 N_K u_{\max})$, where $u_{\max}$ is the parameter value at which the singularity P is encountered. Knowing that p is bounded, we can now show that its derivatives are continuous in u and v from the equation obtained by integrating (5) once:

$$p_{,v} = \int_0^u du' K(u',v)e^{p(u',v)}$$

and similarly for $p_{,u}$.

The estimates for ρ proceed similarly, except that here we do not have the option of making the variable constant on two geodesics. Instead we first show that the derivatives of ρ are bounded on the geodesics κ and γ_0 by using

$$L_{\alpha\beta}\dot{\kappa}^\alpha \dot{\kappa}^\beta = -(H/\rho)\frac{d(\rho_{,v}/H)}{dv}$$

(which comes from a direct calculation of the Riemann tensor) and then integrating and using the bounds already assumed on the integral of the left hand side. Then the estimate proceeds as before with (5) replaced by (4).

In this way one can show that in null coordinates the metric coefficients and their first derivatives are continuous and uniformly bounded in a neighbourhood of the singularity.

One can now use standard techniques to extend the functions H and ρ in the (u, v)-plane. First note that the domain U covered by the coordinate-image of V, defined by $0 < v < \phi(u)$, $0 < u < u_0$, satisfies a cone-condition: specifically, if (u, v) is in the singular boundary

$$\partial^+ U : v = \phi(u),$$

with $|v| < \epsilon/2$ then for

$$|(u - u', v - v')| < \epsilon/2$$
$$u - u' + v - v' > 2|u - u' - v + v'|$$

we have $(u', v') \in U$.

Thus to construct an extension we take a smooth nonnegative function $\chi : \mathbb{R}^2 \supset S^1 \to \mathbb{R}$ with support on the set corresponding to the above cone,

$$\{(u', v') \in S^1| - (u' + v') > 2|v' - u'|\}$$

and normalised to

$$\int_0^{2\pi} \chi(\cos\theta, \sin\theta)d\theta = 1,$$

and let ψ be a smooth function equal to 1 on $\partial^+ U$ with support lying within $\epsilon/2$ of $\partial^+ U$.

Then, putting $x^1 = u$, $x^2 = v$, define

$$\bar{H}(x^1, x^2) = \int_U \chi\left(\frac{y - x}{|x - y|}\right) \frac{(x^i - y^i)\,(H(y)\psi(y))_{,i}}{|x - y|^2} d^2y$$
$$= \int_{S^1} H(y)\chi(\omega)d\omega$$

with $\omega = (y - x)/|x - y|$, and similarly for ρ. It is clear from the above formulae that this is indeed a C^{1-} extension.

We stress that, like the other extensions considered in this chapter, the operation has been purely geometrical, with no reference to field equations or to the physical interpretation of the Ricci tensor. Thus when, as here, the Ricci tensor is discontinuous, there is no guarantee that, if the Ricci tensor of the original space-time corresponds to a positive energy, then the distributional Ricci tensor of the extension has the same property. This is because there is no control at all on the distributional second derivatives. It would be of interest to know whether, in the case of spherical symmetry, such positivity conditions could be maintained if one imposes the additional requirement that the function ϕ specifying the boundary is smooth (C^2), in which case an extension can be constructed that does preserve this second derivative. It would appear that even in this case further continuity conditions are required on the integrals of the Riemann tensor.

7.2 The future boundary of a globally hyperbolic space-time

When the space-time is globally hyperbolic, the possibility exists of making an extension through the whole of that part of the future b-boundary that is not singular. Note that this is stronger than being able to extend through any given point of that part, which is usually the best that can be done. (Little work has been done on the general problem of understanding the situation where there are boundary points that are non-singular, in the sense that each can be extended through, but where there is no simultaneous extension through all of them.) As an illustration of the techniques involved, we shall consider the case of C^{2-} metrics. Here the Riemann tensor is locally bounded, so a singularity is heralded by the Riemann tensor being unbounded. While this is the simplest case, it has the disadvantage (as compared to the Sobolev or Hölder cases) that it is not possible to control the size of the Riemann tensor in an extension. Where the Riemann tensor is bounded, the results of chapter 4 allow us to construct coordinate systems in which the metric is C^{1-} up to the boundary, and so a C^{1-} extension is obtained.

A condition is required to exclude algebraically specialised Weyl tensors. This has two roles: first, it allows us to deduce past-simplicity from Theorem 6.4.2; second it controls the way in which extensions through different points of the boundary relate to each other, enabling us to extend through all the points.

The work in this section was carried out jointly with J. Isenberg.

Theorem 7.2.1
*Let M be a globally hyperbolic space-time with a C^3 atlas **A** in*

which g is C^{2-}. Define $\partial^+_{reg}M$ to be the set of points $p \in \partial_b M$ for which

1. *there is a future-directed causal curve γ ending at p;*
2. *p is a point of generic curvature; and for some (and hence any) frame $\mathbf{E}$ at p, $\mathcal{R}^*(\mathbf{E})$ is bounded.*

Then there is a isometry $\theta : M \to M'$, with (M', g') a C^{1-} space-time, such that $\bar{\theta}(\partial^+_{reg}M) \subset M'$ (where $\bar{\theta}$ is the extension of θ to $\bar{M}$ (Schmidt, 1971)).

Proof

From Theorem 6.4.2 and Proposition 6.3.2, each p in $\partial^+_{reg}M$ is a past-hyperbolic, past-simple, quasi-regular singularity. From the results of 4.7 and 4.8 we can find coordinates covering a neighbourhood of such a p in $\bar{M}$ in which the metric and its first derivatives are bounded (the bound being given by the size of the Riemann tensor). By using smoothing operations in this atlas, together with the results of (Clarke, 1975), we can ensure the existence and uniqueness of timelike geodesics and the applicability of the results of the preceding chapters, despite the low differentiability of the metric.

We now apply the A-boundary construction of 3.5. to the atlas formed by applying a linear transformation to each distance-coordinate chart so as to make the metric take Minkowski form at one point. Condition (2) of 3.5.2 then follows from global hyperbolicity, while condition (3) follows from the non-specialization of the Weyl tensor and Proposition 6.4.1. The result then follows from 3.5.5. $\qquad\square$

7.3 Singularity theorem

We finish by bringing together the arguments of the preceding chapters to formulate what is still the only general theorem about the strengths of singularities.

Theorem 7.3.1 *For a $C^{0,\alpha}$ space-time the following conditions are incompatible:*

1. *M has no holes;*

2. M has no primordial singularity;

3. every singularity p accessible on a future directed causal curve is quasi-regular, a point of generic curvature, and any frame over p has a neighbourhood V in which

$$|R^\alpha{}_{\beta\gamma\delta}(\mathbf{E}) - R^\alpha{}_{\beta\gamma\delta}(\mathbf{F})| < A|\mathbf{E} - \mathbf{F}|$$

for some constant A.

Proof

We have essentially shown this already at the end of 6.5, combined with Theorem 6.3.4 and Theorem 6.4.2. The only condition to be verified is that the Hölder condition in 5.4 can be satisfied in the set K of 6.5. But this is precisely what is ensured by condition (3), when applied to the distance-function of 4.7, used in K. $\square$

7.4 Conclusions

It is perhaps one of the disappointments of the subject of space-time singularities, as it was established by the definitive work of Hawking and Penrose in the early 1970s, that no final satisfactory statement emerges of what a singularity really is like. Instead we have a catalogue of possible ways in which a space-time could break down: it could contain a causally inexplicable hole (6.5); it could have been created with a primordial singularity (6.3); or there could be a breakdown of the smoothness or genericness of the Riemann tensor at the singularity (condition (3) above). For globally hyperbolic space-times (Theorem 7.2.1) this breakdown can be extended to unboundedness of the Riemann tensor (providing that we are prepared to accept C^{1-} extensions), and for spherically symmetric space-times to non-integrability of the Riemann tensor (7.1.2).

But this knowledge only serves to place us at the start of the real problem of general relativity: given, say, a collapsing star, what actually happens? Is there, for instance, a solution which, though formally singular in the classical sense of not admitting a C^{1-} metric, still satisfies Einstein's equations distributionally? Or must one accept that the Riemann tensor is so badly behaved that there is no hope of making any sensible extension at all? The answers to these questions involves detailed considerations of distributional solutions to Einstein's equations, leading into an area that is only now starting to be explored, and taking us well beyond the methods of this book.

8

Singularity strengths and censorship

In section 6.2.5 we introduced the strong cosmic censorship hypothesis, which implies that singularities other than the big bang would be unobservable. If this were literally true, then it might be thought that the considerations of this book were physically irrelevant. We shall see, however, that the situation is more complex than this. The main thrust of this book has been the attempt to establish a relation between the curvature strength of singularities (in the sense of section 6.1) and their 'genuineness' – i.e. whether or not there is an extension through them. The arguments for cosmic censorship suggest that only sufficiently strong singularities might be censored, and so the crucial question becomes, whether or not the censored singularities are precisely the genuine ones. In view of the importance of this to the whole study of singularities, I give here a more extended account of the cosmic censorship hypothesis.

8.1 The weak hypothesis

The strong cosmic censorship hypothesis was preceded by the weak cosmic censorship hypothesis, first formulated by Penrose (1969), who asked: "does there exist a 'cosmic censor' who forbids the appearance of naked singularities, clothing each one in an absolute event horizon?"

This last term was subsequently given more precision in terms of future null infinity $\mathcal{J}^+$. A full account of this would take us well beyond the scope of this book. In outline, however, $\mathcal{J}^+$ is a boundary attached to a conformal extension of a given space-time (M, g). That is, one supposes the existence of a space-ime $(\bar{M}, \bar{g})$ and a map $\theta : M \to \bar{M}$ such that $\theta^* \bar{g} = (\Omega \circ \theta)^2 g$ for a function $\Omega : \bar{M} \to \mathbb{R}$ with (writing $\mathcal{J}^+$ for the boundary of $\theta(M)$)

154

1. $\Omega = 0$ and $d\Omega \neq 0$ on $\mathscr{I}^+$;

2. Space-time geometry on $\mathscr{I}^+$ is equivalent to that which obtains on the future boundary of the standard conformal extension of Minkowski space.

Clearly, the precise meaning of $\mathscr{I}^+$ depends on what is meant by "geometry" (and hence by "equivalent") in 2. This is a debated issue of considerable delicacy. All versions of the definition, however, can be regarded as asserting the existence of a future null infinity, the definitions differing in the asymptotic properties that they require space-time to have in a neighbourhood of this infinity.

A space-time having a both $\mathscr{I}^+$ and its past analogue $\mathscr{I}^-$ is termed weakly asymptotically simple and empty. Both $\mathscr{I}^+$ and $\mathscr{I}^-$ are null surfaces, whose generators are complete if one uses a conformal factor Ω in which the expansion of these generators is zero (as can always be arranged). The notation used here is the reverse of that of Penrose, who interchanges the barred and unbarred quantities. However, in what follows (since there is no detailed discussion of $\mathscr{I}^+$) I will tend to identify M with its image in $\bar{M}$, and so not use bars in discussing these spaces.

An absolute event horizon means, in this context, the boundary of the past of $\mathscr{I}^+$. In other words, it is the boundary of those events from which it is not possible to escape to future null infinity. The weak hypothesis thus asserts that, though singularities other than the big bang may be observable, any observer of a singularity cannot subsequently escape to the outer region defined by $I^-(\mathscr{I}^+)$.

The physical justification for the hypothesis is the idea that singularities should be associated with high energy densities, which are in turn normally associated with the focussing of null congruences. This suggests that singularities may be contained in trapped surfaces (surfaces such that both congruences of null geodesics leaving them are converging), which – subject to other global conditions – are known, as a result of the singularity theorems, to be contained inside event horizons.

Although this has some plausibility, there are certainly many counter-examples to it as it stands, beginning with the work of Yodzis et al. (1973,4). It is generally agreed, therefore, that the hypothesis, if true, will only hold for 'generic' space-times. To formulate this idea, let us restrict attention to weakly asymptotically simple and empty space-times having a partial Cauchy surface S such that all past-directed generators of $\mathscr{I}^+$ enter and remain in the boundary of the future domain of dependence of S. We will also suppose that S is $\mathbb{R}^3$ (in the light of Newman (1989a) this is a real restriction, but not too drastic a one) with some map $\mathbb{R}^3 \to S$ specified in each case so that data on S can be pulled back to data on $\mathbb{R}^3$. Then we can (not entirely consistently) call a set of such things *generic* if the corresponding collection of data-sets on $\mathbb{R}^3$ is open and dense in some suitable topology. We also require that the domain of dependence of S should be the maximal Cauchy development of S, in order to exclude space-times with bits cut out of them.

Then we can annunciate the following:

Weak cosmic censorship. For a generic set of space-times and partial Cauchy surfaces as above, the whole of $\mathscr{I}^+$ lies in the boundary of the future domain of dependence of S.

The last clause is what Hawking has termed the *future asymptotically predictable* condition. This formulation, while perhaps on the right lines, still lacks precision. We have not specified what equations are in force: vacuum, fields, fluids, or kinetic theory; we have not stated what constitutes a 'solution' (and hence a development of data) of these equations, in terms of its allowed differentiability; and there are subtle points revealed by Newman's work (1989a) about what constitutes an asymptotically flat space-time, in terms of causality restrictions.

It has to be said that progress towards proving the above proposition has been limited. The difficulty, in essence, is that we are concerned with a global existence proof for the non-linear equations of general relativity, and such global proofs are notoriously hard to come by: there are some useful non-existence theorems (Rendall, 1992) and existence theorems in special cases (Chrusciel et al, 1989), but little beyond this. It was as a result of these difficulties that many workers have concentrated on the strong hypothesis of section 6.2.5.

8.2 Strong curvature and extension

In an effort to turn into a theorem the argument, given earlier, that singularities should develop inside horizons, the idea was introduced of a 'strong singularity'. It was hoped to define this in such a way that

1. all real singularities were strong singularities, because it should be possible to extend the space-time through anything weaker than these;
2. strong singularities are always inside horizons.

The pursuit of this strategy led to a greatly improved understanding of the nature of singularities, but the problem has shown little sign of yielding to this approach, because of the gap between definitions of strong singularities that work for (1) and those that work for (2). In order to achieve goal (2) we need a definition of strength that is related to focussing, which is achieved by the following alternative definitions of *strong curvature*. They both refer to a timelike (resp. null) geodesic $\gamma : [0, a) \to M$, the singularity being approached as the affine parameter s tends to a, and are conveniently expressed in terms of Jacobi fields that vanish at a point $\gamma(b)$: we define $J_b(\gamma)$ for $b \in [0, a)$ to be the set of maps $Z : [b, a) \to TM$ such that

$$Z(s) \in T_{\gamma(s)}M, \quad Z(b) = 0, \quad \frac{\mathrm{D}^2 Z}{\mathrm{d}s^2} = R(\dot{\gamma}, Z)\dot{\gamma}, \quad \left.\frac{\mathrm{D}Z}{\mathrm{d}s}\right|_b .\dot{\gamma}(b) = 0.$$

For a timelike (resp. null) geodesic we use three (resp. two) such fields independent of each other (and of $\dot{\gamma}$). Their exterior product defines a spacelike volume (resp. area) element, whose magnitude at the affine parameter value s we denote by $V(s)$. The strong curvature conditions are then

SCSK*(Krolak, 1987) For all* $b \in [0, a)$ *and all independent fields* $Z_1, Z_2, Z_3 \in J_b$ *(resp.* $Z_1, Z_2 \in J_b$*) there exists* $c \in$ $[b, a)$ *with* $\mathrm{d}V(c)/\mathrm{d}s < 0$.

SCST*(Tipler, 1977) For all* $b \in [0, a)$ *and all independent fields* $Z_1, Z_2, Z_3 \in J_b$ *(resp.* $Z_1, Z_2 \in J_b$*) we have*

$$\liminf_{s \to a} V(s) = 0.$$

We shall retain the oldest terminology of *strong curvature* for SCST, but follow Newman (1986) in refering to SCSK as the *limiting focussing condition*.

With either of these definitions it seems equally difficult to prove (1) or (2) (which maybe suggests that the line is being drawn in the right place!) Some progress can be made in proving that the singularities corresponding to these definitions of strong curvature are censored (step (1)), but only with rather strong additional assumptions (Krolak, 1987, 1992) that are hard to justify a priori.

8.2.1 Curvature conditions

In the case of (2), in order to prove that weaker singularities can be removed by extension it is necessary to convert the conditions into properties of the Riemann tensor. We have already deduced in section 2.1 conditions for focussing in terms of pointwise bounds on the Riemann tensor. Similar techniques can then be applied to give the following integral conditions. Necessary conditions for SCST are:

Proposition 8.2.1

For both the timelike and the null cases, if condition SCST is satisfied, then for some component $R^i{}_{4j4}$ of the Riemann tensor in a parallely propagated frame the integral

$$I^i{}_j(v) = \int_0^v dv' \int_0^{v'} dv'' |R^i{}_{4j4}(v'')|$$

does not converse as $v \to a$.

In the null case it is possible to give a more refined form of the proposition by considering the Weyl and the Ricci parts of the curvature separately:

Proposition 8.2.2

If $\gamma(v)$ is a null geodesic and condition SCST is satisfied then either the integral

$$K(v) = \int_0^v dv' \int_0^{v'} dv'' R_{44}(v'')$$

or the integral

$$L^m{}_n(v) = \int_0^v dv' \int_0^{v'} dv'' \left(\int_0^{v''} dv''' |C^m{}_{4n4}(v''')| \right)^2$$

for some m, n does not converge as $v \to a$.

A sufficient condition is as follows:

Proposition 8.2.3 *For both the timelike and null cases, if the integral $K(v)$ diverges and has a positive integrand, then SCST is satisfied.*

Similar conditions hold for SCSK but with one less integral in all cases. Although proofs have been given in Clarke and Krolak (1985), there a great many misprints in this paper, and so it will be useful to give correct proofs again here.

We begin with preliminary results from which the others can be deduced.

Lemma 8.2.4
Suppose $f : [0, a) \to \mathbb{R}$ is integrable and satisfies

$$f \geq 0, \qquad \int_0^a dv' \int_0^{v'} dv'' f(v'') < \infty. \qquad (1)$$

Then for all $\eta > 0$ there exists $c(\eta) \in [0, a)$ such that, for all b, v with $c \leq b \leq v < a$ we have

$$Q(b, v) := \int_b^v dv' \int_b^{v'} dv'' f(v'') < \eta.$$

Proof
By integration by parts

$$Q(b, v) = \int_b^v dv'(v - v') f(v').$$

Hence we conclude that

$$Q(0, v) - Q(0, b) = Q(b, v) + (v - b) \int_0^b f(v') dv'.$$

Since the last term is positive, this gives

$$Q(b, v) \leq Q(0, v) - Q(0, b) = \int_b^v dv' \int_0^{v'} dv'' f(v'').$$

But now, by Cauchy's principle of convergence. the convergence in (1) imples that, for c sufficiently close to a, and for all $b > c$, this integral will be less that η, thus proving the lemma.

□

In the next lemma, M_r denotes the set of real $r \times r$ matrices, $P : M_l \to M_k$ is continuous and homogeneous of degree λ, $A : [0, a) \to M_k$ and $B : [0, a) \to M_l$ are functions integrable on compact sets, and $q : \mathbb{R}^k \to \mathbb{R}$ is a positive definite quadratic form. The mapping norm for matrices and the Euclidean norm for vectors are both denoted by $\| \ \|$. We consider the equations (for $Y : [0, a) \to \mathbb{R}^k$, $\Sigma : [0, a) \to M_l$)

$$\frac{\mathrm{d}^2 Y(v)}{\mathrm{d}v^2} = A(v)Y(v) + P(\Sigma)Y \tag{2}$$

$$\frac{\mathrm{d}}{\mathrm{d}v} q(Y(v))\Sigma(v) = q(Y(v))B \tag{3}$$

with initial conditions

$$\left.\frac{\mathrm{d}Y}{\mathrm{d}v}\right|_b = Y_b' \tag{4}$$

$$Y(b) = 0, \qquad \Sigma(b) = 0 \tag{5}$$

imposed at some b in $[0, a)$.

Lemma 8.2.5

Suppose that

$$\int_0^a \mathrm{d}v \int_0^v \mathrm{d}v' \|A(v')\| < \infty, \tag{6}$$

$$\int_0^a \mathrm{d}v \int_0^v \mathrm{d}v' \left(\int_0^{v'} \mathrm{d}v'' \|B(v'')\| \right)^\lambda < \infty. \tag{7}$$

Then there exists a $b \in [0, a)$ such that, for any $Y_b' \in \mathbb{R}^k$ with $\|Y_b'\| = 1$, the solution Y, Σ to (2)–(5) satisfies

$$\|Y(v) - (v - b)Y_b'\| \leq \frac{1}{2}(v - b) \tag{8}$$

for all $v \in [b, a)$.

Proof

From lemma 4, given any positive constants α and β we can choose b so that

$$\int_b^a dv \int_b^v dv' \|A(v')\| < \alpha, \tag{9}$$

$$\int_b^a dv \int_b^v dv' \left(\int_0^{v'} dv'' \|B(v'')\| \right)^\lambda < \beta. \tag{10}$$

From (4) we know that there exists a $\delta > 0$ such that

$$\|Y(b+\delta) - \delta Y_b'\| \le \delta/2.$$

So, if we define

$$K := \sup\{v \mid b \le v < a \text{ and } \|Y(v) - (v-b)Y_b'\| \le \tfrac{1}{2}(v-b)\} \tag{11}$$

then we know that $K \ge b + \delta > b$. The required result (8) is equivalent to the assertion that $K = a$; so we proceed to prove this by contradiction – we assume that $K < a$, in which case we would have equality in the condition in (11) at K (for if not, it would be possible to proceed further than K). In other words

$$\|Y(K) - (K-b)Y_b'\| = \frac{1}{2}(K-b). \tag{12}$$

We now proceed to get a contradiction to (12). From (11) we have, for $b < v < K$, remembering that $\|Y_b'\| = 1$,

$$\frac{1}{2}(v-b) \le \|Y(v)\| \le \frac{3}{2}(v-b). \tag{13}$$

From (3) and (5) we have

$$\Sigma(v) = \int_b^v \frac{q(Y(v'))}{q(Y(v))} B(v')dv'$$

whence, taking moduli and using (13) to estimate the integrand, for $v < K$,

$$\|\Sigma(v)\| \le K_1 \int_b^v \|B(v')\|dv',$$

with $K_1 = 9 \sup(q(x)/\|x\|^2)/\inf(q(x)/\|x\|^2)$. Using this and (13) to estimate the right hand side of (2) we obtain, on integrating (2),

$$\left\| \frac{dY}{dv}(v) - Y_b' \right\| \le \frac{3}{2}(v-b) \left[\int_b^v \|A(v')\|dv' \right.$$

$$+ K_2 \int_b^v \left(\int_b^{v'} \| B(v'') \| dv'' \right)^\lambda dv' \Bigg]$$

where $K_2 = K_1^\lambda \sup(\| P(x) \| / \| x \|^\lambda)$.

Integrating again and using (9) and (10) gives

$$\| Y(v) - (v-b) Y_b' \| \le \frac{3}{2} (v-b) [\alpha + K_2 \beta].$$

If we now choose α and β so that $\alpha + K_2 \beta < 1/3$, we will obtain a contradiction to (12), thus proving the lemma.

□

The final lemma is a recasting of the SCS condition into a more convenient form.

Lemma 8.2.6

The condition SCST implies that, for all $\epsilon > 0$, all $b \in [0,a)$ and all $d \in (b,a)$, there is a Jacobi field $Y \in J_b(\gamma)$ with $\| DY/dv|_b \| = 1$ and $\| Y(v') \| < \epsilon$ for some $v' \in (d,a)$.

Proof

We choose a basis $Z(v), Z(v), Z(v)$ of Jacobi fields in $J_b(\gamma)$,
$_{\ 1}_{\ 2}_{\ 3}$
such that

$$D \underset{i}{Z} / dv |_b = \underset{i}{e},$$

where $\underset{i}{e}$ are orthonormal basis vectors.

Suppose that SCST holds, so that there is a sequence $\{v_i\} \to a$ such that $V(v_i) \to 0$. Recall that the volume $V_i := V(v_i)$ is given by $\det A_i$, where A_i is the matrix with columns formed from the components of the Jacobi fields in a parallely propagated frame:

$$A_i = \left[\underset{1}{Z}(v_i), \underset{2}{Z}(v_i), \underset{3}{Z}(v_i) \right].$$

The characteristic polynomial $\det(\lambda I - A_i)$ must have a root μ_i of magnitude not greater than $|V_i|^{1/3}$, and so there exists a (possibly complex) unit column vector X_i with $A_i X_i = \mu_i X_i$. Introducing the tangent vector Y_i by

$$Y_i(v) = X_i^1 \underset{1}{Z}(v) + X_i^2 \underset{2}{Z}(v) + X_i^2 \underset{3}{Z}(v)$$

we have that $\|Y(v_i)\| = \mu_i \to 0$ $(i \to \infty)$ and $\|DY/dv|_b\| = 1$. Since Y satisfies the Jacobi equation, we have proved the lemma provided that Y is allowed to be complex. But to obtain a real solution we merely note that the real and complex parts of Y separately satisfy the Jacobi equation, and so provide candidates for the statment of the lemma. □

Proof (of Proposition 1)

We prove the timelike case, the null case being essentially identical.

We will show the contrapositive by using lemma 6: namely that if $I^i{}_j$ does converge for every i,j then the conclusion of lemma 6 is false, there being numbers b, d, ϵ such that all Jacobi fields Y in $J_b(\gamma)$ with $dy/dv|_b = 1$ satisfy $\|Y(v)\| > \epsilon$ for all $v \in (d, a)$.

We apply lemma 5 with A having components $R^i{}_{4j4}$ and $P(\Sigma) = 0$ so that (2) is the Jacobi equation. From lemma 5, there is a b such that $\|y - (v - b)Y_0'\| \le (1/2)(v - b)$ for all v in $[b, a)$. This means that $\|Y\| \ge (1/2)(v - b) \ge (1/4)(a - b)$ provided that $v > b + (1/2)(a - b)$. It thus suffices to take $\epsilon = (1/4)(a - b)$ and $d = b + (1/2)(a - b)$ in order to obtain the required contradiction of the conclusions of lemma 6.

□

In order to prove Proposition 2 we need to use the optical equations for null congruences (Hawking and Ellis, 1973, chapter 4). We introduce a new variable $\hat{x}$ related to the area element $\hat{V}$ of such a congruence by $\hat{x}^2 = \hat{V}$ so that the Raychaudhuri equation becomes

$$\frac{d^2 x}{dv^2} = \frac{1}{2}(2\sigma^2 + R_{44}). \tag{14}$$

Here σ^2 is the trace of the square of the shear matrix $\hat{\Sigma}$ which satisfies the propagation equation

$$\frac{d}{dv}\hat{x}^2\hat{\Sigma} = -\hat{x}^2 C \tag{15}$$

where C is a matrix having the components $C^m{}_{4n4}$ of the Weyl tensor in a parallely propagated frame.

If we then rewrite lemma 6, in the null case, in terms of the variable $\hat{x}$, we obtain:

Lemma 8.2.7

The condition SCST implies that for all $b \in [0, a)$ and all solutions $\hat{x}(v)$ of (14) with initial conditions $\hat{x}(b) = 0$ we have that $\liminf |\hat{x}(v)| = 0$.

Proof (of Proposition 2)

This proceeds in exactly the same way as the proof of Proposition 1, using lemma 7 instead of lemma 6, and lemma 5 with $A(v) = -(1/2)R_{44}(v)$, $B = -C$, $P(\Sigma) = (1/2)\mathrm{Tr}\Sigma^2$, $Y(v) = \hat{x}(v)$ and $q(\hat{x}) = \hat{x}^2$ so that equations (2) and (3) become (14) and (15). $\square$

The proofs of the remaining propositions are direct from the optical equations or the corresponding equations for timelike congruences.

8.2.2 Extensions

When it comes to trying to extend the space-time through a putative singularity, one has to cope with the tricky task of extending the metric in such a way that it is sufficiently differentiable to make sense as a solution of the field equations, knowing only bounds on the Riemann tensor, not on the metric itself. We have seen in the previous chapters that, if one tries to construct a purely geometric extension, relying on the properties of geodesics and classical differential geometry, then one needs to maintain a high level of differentiability, and an extension seems only possible in cases where the Riemann tensor satisfies a Hölder condition.

8.2.3 Persistent curvature

The preceding results on curvature conditions show that the connection between the definitions of strong curvature and the behaviour of the Riemann tensor is rather complicated. In order to prove censorship theorems, moreover, what one needs is not so much strong curvature singularities as a means of knowing when conjugate points appear. This led Newman to adopt a different approach, using a different type of definition of strong curvature, called *persistent curvature*.

For a null geodesic γ this is defined as

$$\sup_{I,\underset{\frac{1}{1},\frac{2}{2}}{e,e},m,n} \left\{ \mathrm{mes}(I)^2 \inf_{v \in I} \left| R_{ijkl} \underset{m}{e^i} \underset{n}{e^k} \dot{\gamma}^j \dot{\gamma}^l \right|_{\gamma(v)} \right\}$$

The significance of this condition is that it is precisely what is needed in order to establish conjugate points. Explicitly, one can show (Newman, 1983) that in a space-time satisfying the null convergence condition, a null geodesic that is complete to the future (but not necessarily to the past) admits conjugate points if it is subject to persistent curvature of strength $N = 5.59332\ldots$, where N is the least strictly positive number satisfying $\tan(N^{1/2}) + \tanh(N^{1/2}) = 0$. The principal defect of this definition (from the point of view of the programme enunciated at the start) is that the persistent curvature is a global property of the geodesic, telling one little about the local nature of the singularity at which it ends. It is thus difficult to see how one could link it with information about whether or not the singularity is 'genuine'.

Using this it is possible to prove useful censorship theorems (step (1)), and these provide the most solid general theorems available. As an example (Newman, 1988), we can cite a theorem concerning a naked singularity persisting in the future, in the sense that, however far to the future one draws a cut Σ^- of $\mathscr{I}^-$ and a cut Σ^+ of $\mathscr{I}^+$, one can still find naked singularities visible to their future, in the sense of past-incomplete null geodesics lying in the future of Σ^- and terminating on $\mathscr{I}^+$ to the future of Σ^+. Such a space-time is called *future asymptotically nakedly singular*. If this happens, then some of these persistent naked singularities must be associated with only weak persistent curvature. The statement is as follows:

Theorem 8.2.8

Suppose that space-time satisfies

1. Future asymptotically nakedly singular
2. The null curvature condition
3. The region where strong causality is violated does not extend out to $\mathscr{I}$
4. Weakly asymptotically simple and empty

5. *Globally hyperbolic to the past of some partial cauchy surface.*

Then for any cut Σ^- *of* $\mathscr{I}^-$ *and a cut* Σ^+ *of* $\mathscr{I}^+$, *there exists a two-parameter family of past-endless, past incomplete, affine null geodesics to the future of* Σ^-, *terminating to the future of* Σ^+ *on* $\mathscr{I}^+$, *with each member of the family subject to persistent curvature of strength not greater than* N.

One important feature revealed by this analysis is that the results are obtained only under global asymptotic causal conditions on the space-time that are stronger than those normally imposed, while being in themselves very natural.

8.3 Counterexamples

Considerable progress has been made in finding counterexamples to stronger forms of the censorship hypotheses, thus exploring the nature of the conditions that need to be imposed if the proposition is to be true. All of the interesting examples are non-vacuum, which might suggest that the hypothesis is more likely to be true for vacuum solutions. Inevitably, the examples have high symmetry and this is widely thought to be crucial.

In addition, most of the counterexamples (with the notable exception of those of Ori and Piran, 1988) involve a type of free-particle matter. We turn to this next, to discuss the general formulation of matter equations of this kind.

8.3.1 Free particle matter

By this, I mean a situation where the matter is described in kinetic theory terms as a cloud of particles, all moving (without collision or other non-gravitational interaction) along geodesics. In general, within a small volume about any point in space-time there will be particles having a variety of momenta, the whole situation being describable by a particle distribution function f giving the density of particles in the phase-space P of all momenta. By normalising the momenta to have unit norm (and assuming the particles are identical) we can identify P with the unit cotangent bundle. Then f is governed by the equation of conservation of particles

$$Xf = 0 \qquad (16)$$

(the Vlasov equation), where X is the geodesic flow in P. In other words, this is Liouville's theorem on phase space.

Most of the counterexamples, however, use the singular special case in which, at any given point of space-time, all the particles have the same momentum. This is referred to as *dust*, or *null dust* if the momentum vector is everywhere null (which requires, of course, the replacement of unit cotangent bundle by the null cotangent bundle). In terms of the Vlasov picture, the distribution function in this case collapses to a generalised function concentrated on a 4-surface in P.

When trying to construct extensions of the space-time, there are crucial differences between the regular case (where f is a smooth function) and dust, since global existence results have recently been obtained (Rendal, 1992) for the system of the coupled Einstein and Vlasov equations in the regular case. On the other hand, if f is (as we need for dust) a generalised function, then the metric must also be regarded as a generalised function (though one that may coincide with a continuous or C^1 function) and the same applies to the connection and hence to X. Thus the Vlasov equation formally involves a product of generalised functions.

If we have a simple dust, then the dynamics can be handled by imposing the form $T_{ij} = \rho u_i u_j$ on the energy momentum tensor, where u is the unit velocity vector. The Einstein equations then give a conservation equation for T which in turn implies that u is tangent to a congruence of geodesics, thus determining the system completely. If one tries to extend the soution, however, through a singularity at which adjacent shells of matter may be crossing, then any extension is liable to be a *multidust* model, in which several dust flows are superimposed at each point. In this case the geodesics equations of motion of the particles ar no longer implied by Einstein's equations, but have to be imposed separately, while the Vlasov equation (which does this for non-singular distributions) sufferes from the difficulties already described.

There are two ways of describing the situation in this case.

(i) *6-form formulation.* The distribution function f specifies the particle density in the sense that if U is a compact subsurface of P with boundary such that πU is a spacelike submanifold Σ with boundary of space-time M, then the number of particles in Σ having momenta in U is given by $\int_U f X \rfloor \omega$ where ω is the canonical volume form on P. Thus the form

$$\nu = f X \rfloor \omega \tag{17}$$

is equivalent to f and directly specifies particle fluxes over hyper-surfaces. Conservation of particles is then

$$d\nu = 0 \tag{18}$$

(from which the Vlasov equation follows on writing

$$d(X \rfloor f\omega) = \mathcal{L}_X(\omega f) = f\mathcal{L}_X\omega + (Xf)\omega$$

and noting that $\mathcal{L}_X\omega = 0$.) If X is smooth, then (17) is equivalent to

$$X \rfloor \nu = 0 \tag{19}$$

i.e. (16) is equivalent to (18) and (19) via the identification (17). We can generalise ν to be a current of degree 6 (using de Rham's terminology: a current of degree p is a generalised p-form, dual to test-$(n-p)$-forms on a manifold of dimension n).

(ii) *Hypersurface formulation.* An alternative to the above is to follow the procedures established in the Newtonian case of describing the dust flow by Lagrangian coordinates in which the particles are labeled by coordinates x^α ($\alpha = 1\ldots3$), and their world lines are parametrised either by proper time s or by a global time coordinate t. If $X(s, x^\alpha)$ is the tangent vector to the particle with parameters x^α at proper time s, and $u(s, x^\alpha)$ the corresponding covector, then the map $\phi : (s, x^\alpha) \mapsto u(s, x^\alpha)$ takes a neighbourhood of $\mathbb{R}^4$ into a set N in the unit cotangent bundle P. In favourable circumstances ϕ can be an embedding, but the arguments given above concerning the Vlasov equation show that it cannot be smooth. If ϕ is C^1 (for which there is evidence), then one can easily relate this to the Vlasov setting, as follows.

The 6-form ν will in this case be concentrated on the surface N; more specifically, it has the form

$$\nu = N^c \wedge n \tag{20}$$

where N^c is the current of degree 3 (generalised 3-form) corresponding to N, defined by $\langle N^c, \theta \rangle = \int_N \theta$ for all test 4-forms θ, and n is a nowhere-zero 3-form which can be regarded as being defined on N – i.e. only the value of i^*n, where i is the embedding of N in P, enters into the value of ν. Then (18) implies (if N is without boundary)

$$dn = 0 \qquad (21)$$

while if X is smooth (19) implies (since n is nowhere zero) that

$$X \text{ is tangent to } N$$

and $X \rfloor n = 0.$ \qquad (22)

To summarise: *the conserved particle distribution current ν in P is specified by a nowhere zero 3-form n satisfying (21) and (22). If the geodesic flow is smooth, these are equivalent to (18) and (19) (and hence to the Vlasov equation) via (20).*

The formulation based on (21) and (22) was used in Clarke and O'Donnell (1993) to establish evidence for multidust extensions through some naked singularities with dust.

8.3.2 *Tolman-Bondi models*

The spherically symmetric solutions for dust were found by Tolman (1934) and Bondi (1947). At some time specified as 'initial' the density of matter and its velocity, as functions of a radial coordinate (say the areal radius R), can be specified arbitrarily; the matter can be thought of as disposed in spherical 'shells' labelled by a comoving coordinate r, which can be taken as the value of R at the initial time. For a certain stable set of initial conditions it turns out that the analytic form of the solution breaks down at a stage where dR/dr becomes zero: infinitessimally neighbouring shells approach each other. This is termed a *shell crossing singularity*. We shall follow recent authors in reserving the term for the case where this happens only at isolated and non-zero values of R, calling the other cases *shell focussing singularities*. Shell crossing singularities were first found by Yodzis et al. (1973,4) and have been investigated extensively since. The shell focussing singularity that can be formed at the centre satisfy the limiting focussing condition (Newman, 1986) and even, provided the initial distribution is not smooth, the strong curvature condition (Gorini et al, 1989). With appropriate conditions the examples violate both strong and weak cosmic censorship.

If this is to be accepted as a true counterexample to cosmic censorship, then we need to verify that it is generic (i.e. the singularity does not disappear, nor does a horizon form, when it is perturbed) and that the singularity is genuine, so that there is no extension possible through it.

Concerning the first, there is now evidence that perturbation within the wider class of axially symmetry do not change the situation, and perturbations from pure dust to a particle gas (satisfying the Einstein Vlasov equations, but with zero initial internal energy). Shapiro and Teukolsky (1991) have recently presented a numerical example of a cloud of non-colliding particles (Boltzman gas) with only axial symmetry which collapses to a spindle-like singularity that appears to be naked. Though we have a relaxation both of the symmetry and of the nature of the matter, the procedure has been criticised: Wald and Iyer (1991) have pointed out that nakedness has not been proven rigorously, and Rendal (1992) has drawn attention to analogous Newtonian situations where the initial conditions used by Shapiro and Teukolsky (zero temperature) are an exceptional case.

Further evidence comes from the work of Ori and Piran (1988) who find naked singularities in a collapse of a perfect fluid with a barytropic equation of state. This solution has even more symmetry, being self-similar as well as spherically symmetric, but other examples (Lake, 1991) suggest that this is not the reason for the singularity.

Perhaps more significant is the widely believed idea that the shell crossing singularities are not true singularities, but admit extensions satisfying Einstein's equations in a distributional sense. Evidence for this has been presented in Clarke and O'Donnell (1992). Since shell crossing singularities do not satisfy even the limiting focussing condition, it might be thought that this approach would fail with the stronger shell focussing singularities; Papapetrou and Hamoui (1967) have, however, constructed an extension through a simple shell focussing singularity which can be shown to satisfy Einstein's equations, suggesting that this approach may be generally applicable.

References

Adams R.A. (1975), *Sobolev Spaces*, Academic Press, New York

Beem J.K. and Ehrlich P.E. (1971), *Global Lorentzian Geometry*, M. Dekker, New York etc.

Birkhoff G. (1937), Moore-Smith convergence in general topology, *Ann. Math.*(2) **38**, 39–56

Bondi H. (1947), Spherically symmetrical models in general relativity, *Mon. Not. Roy. Astron. Soc.* **107**, 410—425

Bosshart B. (1976), On the b-boundary of the closed Friedmann model, *Comm. math. Phys.* **46**, 263–268

Calderon A.P. and Zygmund A (1952) On the Existence of Certain Singular Integrals, *Acta Math.* **88**, 85–139; (1956) On Singular Integrals, *Amer. J. Math.* **78**, 289–309

Chrusciel P.T., Isenberg J. and Moncrief V. (1990) Strong Cosmic Censorship in Polarized Gowdy Spacetimes, *Class. Quantum Grav.* **7**, 1671—1680

Clarke C.J.S. (1978), The singular holonomy group, *Comm. math. Phys.*, **58**, 291–297

Clarke C.J.S. (1982), Space-times of low differentiability and singularities, *J. Math. Anal. Applic.*, **80**, 270–305

Clarke C.J.S. and Krolak A. (1985), Conditions for the occurrence of strong curvature singularities, *J. Geom. Phys.* **24**, 127–143

Clarke C.J.S. and O'Donnell N. (1993) Dynamical extension through a space-time singularity, *Rendiconti del seminario matematico, Università e Politecnico Torino* **50**,1 (1992), 39–60

Dodson C.T.J. (1980), *Categories, Bundles and Spacetime Topology*, Shiva, Orpington

Einstein A. (1918), Kritisches zu einer von Herrn de Sitter gegebenen Lösung der Gravitationsgleichungen, *Sb. Preuss. Akad. der Wiss.*, 1918 (1), 448–72

Eddington A. S. (1924). *Nature*, **113**, 192

Geroch R.P., Kronheimer E. H. and Penrose R. (1972), Ideal points in general relativity, *Proc. Roy. Soc. Lond.*, **A 327**, 545–567

Geroch R.P. and Traschen J. (1987) *Phys. Rev. D* **36**, 1017–31

Gorini V., Grillo G. and Pelizza M. (1989) Cosmic Censorship and Tolman-Bondi Spacetimes *Phys. Lett. A* **135** 154—158

Hawking S.W. and Ellis G.F.R. (1973), *The large-scale structure of space-time.* Cambridge University Press.

Hocking J.G. and Young G.S. (1961), *Topology.* Addison-Wesley.

Hughes T.J.R., Kato T. and Marsden J.E. (1977) Well posed quasi-linear 2nd. order hyperbolic systems with applications to non-linear elastodynamics and general relativity, *Arch. Rat. Mech. Anal.* **63**, 273–294

Johnson R. (1977), The bundle boundary in some special cases. *J. Math. Phys.* **18**, 898–902

Kato T. (1975), Quasi linear equations of evolution, with applications to partial differential equations, *Lecture Notes in Mathematics,* Springer, **448**, 25–70

Kelley J.L. (1955), *General topology.* Van Nostrand, New York.

Kobayashi S. and Nomizu K. (1963), *Foundations of Differential Geometry,* Interscience, New York

Krolak A (1987), Towards the Proof of the Cosmic Censorship Hypothesis in Cosmological Space-Times, *J. Math. Phys.* **28** 138—141

Krolak A (1992), Strong Curvature Singularities and Causal Simplicity, *J. Math. Phys.* **33** 701—704

Lake K (1991), Naked Singularities in Gravitational Collapse which is not Self-Similar, *Phys. Rev. D* **43** 1416—1417

Misner C.W. (1967), Taub-NUT space as a counter-example to almost anything. *in Relativity Theory and Astrophysics,* ed. J. Ehlers, (Lectures in applied mathematics, vol 8), American Mathematical Society.

Newman R.P.A.C. (1983), Cosmic censorship and curvature growth, *Ge. Rel. Grav.* **15** 641

Newman R.P.A.C. (1984), A theorem of Cosmic Censorship – a Necessary and Sufficient Condition for Future Asymptotic Predictability, *Gen Rel. Grav.* **16**, 175–192; Persistent curvature and cosmic censorship *Gen Rel. Grav.* **16** 1177—1187

Newman R.P.A.C. (1986), Strengths of Naked Singularities in Tolman-Bondi spacetimes, *Class. Quantum Grav.* **3**, 527–539

Newman, R P A C (1989a) The global structure of simple space times, *Comm. Math. Phys.* **123** 17—52

Newman, R P A C (1989b), Black holes without singularities, *Gen. Rel. Grav.* **21** 981—995

Ori A. and Piran T. (1988), Self-Similar Spherical Gravitational Collapse and the Cosmic Censorship Hypothesis, *Gen. Rel. Grav.* **20**

7—13

Palais R.S. (1968), *Foundations of global non-linear analysis*, Benjamin.

Papapetrou A. and Hamoui A. (1967), *Ann. Inst. H. Poincaré* **9** 343-364

Penrose R. (1969), Gravitational collapse: the role of general relativity, *Riv. del Nuovo Cimento* **1** (Numero Spec.) 252—276

Penrose R. (1974), Gravitational collapse, in *IAU symposium 64 on Gravitational Radiation and Gravitational Collapse* Reidel, Dordrecht, 82—91 *Ann. N Y Acad. Sci.* **224** 125

Rendal A. (1992), On the choice of matter in general relativity, in *Approaches to numerical relativity* ed R A d'Inverno, CUP

Rosen N. (1974), A theory of gravitation, *Ann. Phys.* **84**, 455-473

Shapiro S.L. and Teukolsky S.A. (1991), Formation of Naked Singularities – the Violation of Cosmic Censorship, *Phys. Rev. Lett.* **66** 994—997

Schmidt B. (1971), A new definition of singular points in general relativity, *J. Gen. Rel. Grav.* **1**, 269-280

Schmidt B. (1973), Local b-completeness of spacetimes, *Comm. Math. Phys.* **29**, 49-54

Schoen R. and Yau S.-T. (1983), The existence of a black hole due to condensation of matter, *Comm. Math. Phys.* **90**, 575-579

Seifert H.-J. (1977) Smoothing and Extending Cosmic Time Functions, *Gen. Rel. Grav.* **8**, 815-831

Spivak M. (1965), *Calculus on manifolds*, Benjamin.

Stein E.M. (1970), *Singular integrals and the differentiability properties of functions*, Princeton U. P.

Synge J.L. (1960), *Relativity: the general theory.* North Holland.

Tipler F.J. (1977a), Singularities and Causality Violation, *Annals of Physics*, **108**, 1-36

Tipler F.J. (1977b), Singularities in Conformally flat space-times *Phys. Lett. A* **64**, 8

Tolman R.C. (1934), Effect of inhomogeneity on cosmological models, *Proc. Nat. Acad. Sci.* **20** 169—176

Vickers J.A. (1987), Generalised Cosmic Strings, *Class. Quantum Grav.*, **4**, 1-9

Vickers J.A. (1992) to appear in *Rendiconti del seminario matematico, Universitá e Politecnico di Torino*

Wald R.M. and Iyer V. (1991), Trapped Surfaces in the Schwarzschild Geometry and Cosmic Censorship, *Phys. Rev. D* **44**

Yodzis P., Mueller zum Hagen H. and Seifert H-J. (1973,4) *Comm. Math. Phys.* **34** 135 and **37** 29.

Index

174